Magical Acts, Hypercubes and Pi

Meanderings through science, medicine and mathematics

BEVERLY ORNDORFF

Science Museum of Virginia

Magical Acts, Hypercubes and Pi:

Meanderings through science, medicine and mathematics
by
Beverly Orndorff

Produced by the
Science Museum of Virginia
and the
Science Museum of Virginia Foundation Inc.

First printing May 1999
Second Edition (E-Book) March 2011
Second printing June 2016
ISBN 13 9780-9889886-3-7
ISBN 10 09889886-3-1

.

Dedication

To my wife Barbara, my best friend and valued soul mate.
And to the late Marv Garrette,for his support, inspiration and
goading when I needed it.

Table of Contents

Acknowledgments

The publication of this collection has been made possible by financial support from the *Richmond Times-Dispatch* and the Science Museum of Virginia Foundation, and prime movers behind that support included Dr. Walter R. T. Witschey, the Science Museum's director, and William H. Millsaps, the *Times-Dispatch's* executive editor. To them, to their colleagues and staff and to their institutions I express my sincere thanks.

Additionally, I owe heartfelt appreciation to many other people who have supported and encouraged this project, especially to Kathy Albers of the *Richmond Times-Dispatch* Library staff, who often went beyond the call of duty to help me track down some of my past articles in the newspaper's vast files of clippings.

Also thanks to Dr. Maurice Duke, whose editing and comments were particularly helpful, and to Martin Rhodes, for his help and advice on layout and design matters. I'm also grateful for the interest, encouragement and support given by my former colleagues and bosses at the *Times-Dispatch*. I wish to express, too, my gratitude to the five managing editors under whom I served from 1957 to 1997—the late John H. Colburn, John E. Leard, Alf Goodykoontz, the late Marvin E. Garrette, William H. Millsaps and Louise Seals. Each provided a nurturing climate that ultimately led to this collection.

And most of all, thanks to my wife Barbara for her helpful comments as I prepared my manuscript and for her patient support throughout, and to my late parents, who by example and environment, encouraged a love of reading and learning.

Preface

My first science column for the *Richmond Times-Dispatch* appeared Sunday, January 13, 1963, and thereby began what was to become for me a weekly routine for the next 34 years. That first one was about biological clocks, which influence such events as the flowering of plants and the daily sleep-wake cycles of bees, bats, crabs and people, and it explained how scientists were trying to understand how such clocks work down at the level of the cell. They still are, and the daily rhythms among living things was a topic I revisited numerous times through my years of science reporting. In fact, I revisited many of the topics I wrote about in the early 1960s—space exploration, DNA, organ transplants, the origin of the universe, to name a few. The '60s formed a foundation for much of my subsequent career.

I started writing a science column at the urging of the late John H. Colburn, who was the managing editor of the *Times-Dispatch* from 1949-1963 and who encouraged increased science coverage at the paper following the Soviet Union's techological scoop by placing the world's first artificial satellite, Sputnik, into an Earth orbit on October 4, 1957. I had been a full-time journalist less than four months when Sputnik shook the world, and it turned out to have been an event that shaped my subsequent career at the *Times-Dispatch*, through Mr. Colburn's early influence and encouragement of my interest in science reporting.

Writing a science column through the years was truly a labor of love, rarely a chore. Throughout those decades, my editors gave me considerable freedom to choose column topics, and what I chose were generally subjects that interested me—on the theory that if they interested me, they just may interest readers, too. That, of course, wasn't always true. Sometimes, I'm sure that my personal fascination with certain esoteric topics, such as the intricacies of quantum mechanics, caused many a reader to tune out within the first few sentences and jump to less dense articles on other pages of the newspaper.

I particularly liked topics that illuminated, stimulated or provoked, such as why eggs are shaped as they are, or how living things reflect deep organizing principles in nature, or what the connections are between brain and mind. I liked the quirky, such as an analysis showing how the laws of the universe guarantee that toast will hit the floor butter-side down; the odd, like why a wintergreen Lifesaver emits a blue flash when crunched, and the unusual, such as the story about a famous artist experiencing an outburst of creativity as Alzheimer's disease began stealing his mind. Also, I liked pieces with real-world relevance: Would you keep dryer by walking or running in the rain? What accounts for those 'it's a small world' coincidences that we all experience? What goes on in our bodies when we sneeze? And, I liked to explore the science behind the news, such as a run of bad luck experienced by an airline company, or why eating insects, ala Air Force pilot Scott O'Grady, was a smart thing to do under the circumstances.

My initial model for science column writing back in the early 1960s was the late William L. Laurence of the *New York Times*. He was one of the nation's pioneering newspaper science writers, winner of two Pulitzer Prizes. He was the only journalist present at the first explosion of an atomic bomb, at Almagordo, N.M., and although he knew about the World War II Manhattan Project's efforts to develop such a bomb, he agreed not to write about it because of national security concerns until after it was first used. He won his second Pulitzer, in 1946, for his inside account of the bomb's development during the war years.

Mr. Laurence wrote a weekly science column in the *Times'* Week in Review section that generally featured research reported in a science journal or a broad topic that an expert explored in the monthly publication, *Scientific American*. I had learned early in my science writing career that science exists where scientists communicate—in journals, at meetings. So I began following Mr. Laurence's example by scanning *Scientific American* and journals for science column material, because those were the main meeting places of scientists, the places where the action was—and continues to be, of course. It was a habit that persisted throughout my career.

This collection represents some of the more than 1,600 columns I wrote through the decades, plus a few articles that appeared in the regular news sections of the newspaper. All are essentially as they were originally published, but I have done some editing and minor rewriting, mostly to correct original errors.

In sifting through hundreds and hundreds of my columns, I found that many are now seriously dated, many were clunkers, and many were simply forgettable. I tried to select those with some degree of enduring variety, interest, provocation or amusement.

Foreword

In 1998 the Science Museum took the unusual step of creating an award in Bev's honor, the Beverly Orndorff Award for Exceptional Service to Public Understanding of Science. Furthermore, Bev Orndorff was the first to receive this award named in his honor.

As I said during the presentation, "For 37 years he took complex science subjects and wrote about them so that they could be understood by everyone, not just scientists. For his years of dedication, we're not only naming the award after Bev, we are presenting the first award to him."

What a happy circumstance: Bev Orndorff's lifelong passion is identical to the chief mission of the Science Museum of Virginia. Both work to increase the public understanding of science. It is for this reason that we are delighted to help bring this tasty volume of Bev's columns to the public as a collection.

As a science and medicine writer Orndorff was called "Mr. Wizard" by his colleagues at the Times-Dispatch. It was a nickname of respect from other staffers who saw Bev as a matchless source of information about medicine, science and mathematics. To me he is a master storyteller. Whether his characters are atoms or quarks, planets or meteors, new discoveries or science history, Bev always crafts a clever column that is accurate, precise, informative, understandable and engaging.

Bev retired from the Times-Dispatch after 40 years at the newspaper. His career began in 1957 as a reporter for the paper's Charlottesville bureau. In 1960, he moved to Richmond and began his specialty of science and medicine. His assignments took him into operating rooms, laboratories and science conferences around Virginia and the nation. He covered early organ transplant operations. He wrote about the emerging space program, including the Apollo moon missions. Whether the news was in astronomy, biology and medicine, physics, chemistry or mathematics, Bev was on the job.

In 1957 Bev Orndorff received his bachelor's degree in physics from the University of Virginia. He was the first nonscientist to be given an honorary life membership to the Virginia Academy of Science. He has also been given the American Heart Association's Howard W. Blakeslee Award, the Virginia Museum of Natural History Foundation's William Barton Rogers Individual Award, the Virginia Academy of Science's Ivey S. Lewis Distinguished Service Award, and the American Chemical Society's Virginia Division Distinguished Service Award.

In addition, Orndorff was the first recipient of Duke University's National Association of Science Writers Fellowship, and he has received numerous awards from the Virginia Press Association and other journalism organizations. When he retired, the Media General Foundation—the parent company of the Richmond Times-Dispatch—established a $25,000 Virginia Commonwealth University School of Mass Communications scholarship in his name.

We know you will enjoy these science tales from Bev Orndorff as much as we do!

Walter R. T. Witschey, Ph.D.,
Director, Science Museum of Virginia

LUCK, CHANCE, COINCIDENCE

Probable impossibilities are to be preferred to improbable possibilities.
ARISTOTLE, *Poetics*

The following group of columns reflects a long-time fascination I have with probability and its various manifestations—coincidence, luck, chance, the "law of averages," and randomness. I am convinced that understanding the workings of probability is about as important to one's view of the world as understanding any single subject in science and mathematics, and the implications are as profound and far-reaching as any of the concepts from evolution, quantum mechanics or Newtonian physics. So whenever the opportunity arose to deal with some aspect of probability—neat coincidences, the lottery or highly visible, apparent runs of bad luck—I grabbed it.

Coincidences Are Life's Magical Acts
DEC. 13, 1981

Coincidences are life's magical acts. They astound us, and even though we think that there's really nothing occult about them, sometimes we may wonder.

Consider, for example, a coincidence originally reported in *Life* magazine in 1950 and recounted in the book, *Lady Luck, The Theory of Probability*, by mathematician Warren Weaver.

On March 1, 1950, all 15 members of a church choir in Beatrice, Neb. were late—each for different, ordinary reasons—for the 7:20 choir practice. The church was destroyed by an explosion at 7:25 p.m. Because everybody was late, no one was in the church at the time of the explosion.

Then there's the Apollo 13 near-disaster. Not only was the moon-bound spacecraft labeled Apollo 13, but it was launched at 1:13 p.m.—13:13 in 24-hour time designation—according to the mission control clocks at Houston's Johnson Space Center. And in its three-man crew was the 13th man to go to the moon. On the 13th day of April, an oxygen tank aboard the craft exploded, when it was about 200,000 miles from Earth. The moon landing was aborted, and the crew returned to Earth several days later with the considerable help of mission controllers and others after an extremely close brush with tragedy.

Also, there is a cosmic coincidence that turns out to be responsible for one of the most dramatic of celestial sights, that of a total solar eclipse. A total solar eclipse occurs when the moon is precisely in front of the sun from our point of view, resulting in the moon's disc covering the sun's disc. It turns out that we perceive the apparent sizes of the two discs as being about identical, even though the bodies are of vastly different sizes. The moon is 2,160 miles in diameter, while the sun is 864,000 miles in diameter. But by coincidence, the moon's distance is such that its disc is about the same size as the sun's, from our vantage point, and that coincidence enables it to cover the sun's disc neatly during our views of a total solar eclipse.

On a less dramatic note, there are the kinds of coincidences that make feature stories for newspapers. Last summer, a woman was returned to her room in a Richmond area hospital after an exploratory surgical procedure to find that her roommate was the same person who shared her room at another hospital five years earlier. And the woman's surgical procedure was the same one that her roommate had when they first shared a hospital room.

History, and daily life, teem with coincidences.

Isaac Newton, one of the age's giants in the history of modern science, was born the year that another science luminary, Galileo, died.

Thomas Jefferson and John Adams, both prominent figures in this country's early history, died the same day—July 4, 1826, which incidentally was the 50th anniversary of the Declaration of Independence that Adams signed and Jefferson helped draft.

Mark Twain was born in a year when Halley's Comet was in the sky and died 76 years later, when Halley's Comet was once again in the sky.

In everyday life, practically everyone has a story of a personal coincidence. You may have been thinking of someone when that person calls you. There's the "small-world" class of coincidences in which you find that a newly met person, such as a seat mate on an airplane, is a friend of a friend or relative.

Coincidence has undoubtedly played a role in establishing and reinforcing superstitions, like the one about 13 being an unlucky number. In some cases, coincidences have been taken as evidence of ESP, as in instances of receiving a call or letter from someone being thought about. And in some cases, there is the suggestion that something mysterious is going on that we don't understand, such as the patterns of births and deaths of notable figures.

What is one to make of coincidences? Are they evidence of some hidden phenomena or connections that we don't yet understand, or are they merely long shots that are played out?

The prevailing scientific view is that there's no need to appeal to the occult, the paranormal or "hidden" laws of nature to explain coincidences. They are all the result of what men from Aristotle to present day mathematicians describe as the improbable being more probable than we think. In some cases, according to experts on probability, events are more likely to coincide than our common sense notions tend to suggest.

Take one version of the "same day" kind of coincidence, the birthday coincidence. Almost everyone is aware of a coincidence of birthdays among his or her circle of friends or relatives. At first glance, that seems amazing considering that there are 365 different opportunities for birthdays to occur among a random group of people. Yet, the arithmetic of probability shows there's a better than a 50-50 chance that in any random group of 23 people, two will have the same birthday. If there are 50 people, the chances are 97 out of 100 that two will have the same birthday.

Or take the "small world" problem. Martin Gardner, author of the "Mathematical Games" feature in *Scientific American*, noted some years ago that the world may be smaller than we think. He referred to an experiment in which a person wants to transmit a document to someone else in the nation. The person sends the document to someone, with the instructions he or she is to pass it along to someone else who might know the target person. That someone else has the same instructions.

Most people guess that the chain would involved about 100 people before the document reaches its destination, Gardner said. An experimental psychologist who actually performed the experiment, however, found that the person-links in the chain ranged from two to 10, with a median of five.

Gardner suggests that in the "unthinkably intricate snarls of human history, with billions on billions of events unfolding every second around the globe," some people believe it's amazing that more coincidences aren't publicized. He quotes G.K. Chesterton as saying, "Life is full of a ceaseless shower of small coincidences....It is this that lends a frightful plausibility to all false doctrines and evil fads."

Experts on probability note that even though something may have a small probability of occurring, there is nevertheless the chance it will occur. Weaver made a rough calculation of the Nebraska choir members case that indicated a one in a million chance that various reasons for lateness among them would all occur in one evening. It was a small chance, but it was not a zero chance. Therefore, it could happen.

The fact that events in general have something better than a zero chance of happening, albeit tiny chances in many cases, is involved in an ingenious fun line of reasoning developed by a University of Richmond mathematician a few years ago. Dr. Arthur T. Charlesworth's argument concludes that events are bound to repeat themselves if the human race and the universe exist long enough.

His argument goes like this: If you point a television camera at any particular event or scene, the picture you would see now is bound to be repeated sometime because there are only so many ways that a television image can be produced.

A television picture is produced by the shading of individual little "boxes" running across and down the screen. There is only a finite number of shades that each of those boxes can assume. So, Dr. Charlesworth reasons, there is an extremely small but finite chance that any given image produced by the television camera will eventually be repeated, given sufficient time. In terms of what the camera sees, therefore, history will repeat itself. The numbers are greater and the chances are considerably smaller, but the same line of reasoning applies to the images formed on our retinas. Given enough time, therefore, the coincidental sense involved with *deja vu*, the feeling that one has seen a scene before, is bound to occur, Dr. Charlesworth suggests.

His argument invokes a central lesson from probability theory, namely, given large numbers of events and sufficient time, the improbable is practically bound to occur.

Actually, there are some areas of our lives when coincidence poses problems. In scientific research, especially medical research, investigators have to be constantly vigilant against the possibility that an outcome was due to chance, or coincidence, rather than to the phenomenon being studied. Many diseases, for example, have a natural history of ups and downs, of good periods and bad periods. Was a patient's improvement due to the drug being tested, or a natural remission? Are several cases of leukemia in a neighborhood within a several-year period due to something in the environment, or to an unfortunate coincidence?

Pollsters are plagued by the same sorts of problems. That's why they use terms like "confidence level," which is a measure of the degree to which chance may have been involved in their survey results. A given poll may have a confidence level of 0.05, which means that if the poll were repeated, there's a 5 percent chance the results will be different from those obtained. Or, expressed another way, there's a 95 percent chance that the results will be the same.

Some coincidences may be made dramatically mysterious, and suggestive of "hidden linkages," if there is a careful selection of facts.

Not long ago a newspaper published a number of anecdotes about coincidences, including one about similarities between the Lincoln and Kennedy assassinations. Both presidents were killed on a Friday, next to their wives, by gunshot wounds to the right rear of their skulls. Both were succeeded by men named Johnson and both lost children while in the White House.

The account also said that Lincoln's secretary, named Kennedy, warned him not to go to the theater, where he was killed. Kennedy had a secretary named Lincoln who warned him against going to Dallas, where he was killed. The account says Lincoln's assassin shot the president in a theater and fled to a warehouse, while Kennedy's accused assassin shot from a warehouse and fled to a theater. The account says Lincoln was shot in Ford's Theater, while Kennedy was shot in a Lincoln, a Ford product.

Does all that suggest some sort of connection between the lives of the two men? Are the coincidences too improbable to explain by the laws of probability? If other elements of their lives are examined, a supposed patterns of linkages seems to weaken.

Lincoln's first name was Abraham; Kennedy's, John Fitzgerald. Lincoln was born in Kentucky, the second of two surviving children, to poor parents. Kennedy was born in Massachusetts, the second of nine children, to rich parents. Lincoln was born in February, and Kennedy in May. Lincoln's father's name was Thomas and his mother's name was Nancy. Kennedy's father's name was Joseph and his mother's, Rose. Lincoln's father was a carpenter; Kennedy's, a financier.

Lincoln's wife's name was Mary; Kennedy's, Jacqueline. Lincoln had four sons. Kennedy had a daughter and a son, plus a son who died just a few hours after being born. Lincoln was killed early in the second term of his presidency, while Kennedy was killed during his first term.

And so on. Except for certain, sometimes strained similarities, there appear to be no obvious, occult or mysterious links between the lives of the two men.

Only chance accounts for certain, selected similarities—similarities that Chesterton would undoubtedly include in his "ceaseless shower of small coincidences" that life brings.

'It's a Small World,' We Say, And Probability Proves It
SEPTEMBER 29, 1991

There's nothing like a vacation trip across the United States to reaffirm the truth of the adage, "It's a small world."

People from a small Shenandoah County town in Virginia, for example, recently visited relatives in a small western Wyoming town 2,200 miles away. The son of the Wyoming couple is dating a girl from another Wyoming town, in another part of the state.

It turns out that the Shenandoah County couple has a friend back home who originally came from the same small Wyoming town and who knows the young lady's family. .

Many travelers have small-world stories. On one out-of-state airplane trip, I found that the passenger sitting beside me knew one of my college friends quite well. At a meeting in California, I met someone who knew a friend from the days I lived and worked in Charlottesville.

One doesn't even have to travel to come upon the small-world phenomenon. A long-time colleague here at the *Times-Dispatch* who came from the southwestern part of Virginia, was asked many years ago by his grandfather to look, sometime, for the descendants of a friend of his who had moved to the Central Virginia area in the early 1900s.

It turns out that one of the descendants is my wife—a fact that the colleague and I discovered after working together for 20 or more years.

How is it that the winding paths of people occasionally cross in seemingly unlikely conditions, such as distance, time or circumstance? Are there mystical linkages out there? No, say various mathematicians, it's simply a case of probability in action.

In his book, *Innumeracy*, John Allen Paulos says that such small-world coincidences are quite common. They are common, he said, because each person in the United States has, on average, a fairly large circle of friends and acquaintances, and one's own circle is bound to have some overlap with the circles of many other people.

Martin Gardner, author of *Scientific American*'s "Mathematical Games" column for many years, noted that some social scientists once made a study of the small-world phenomenon. The researchers found, Gardner said in his book, *Aha! Gotcha*, that any two people selected at random in the United States will each know, on average, about 1,000 persons.

The probability that any two people in the nation will know each other, he said, is about one in 100,000.

But the chances that any two people in the country will have one friend in common, according to Gardner, rise to about one in 100. And, he continued, the chances that they are linked by two intermediates are better than 99 in 100.

In other words, if you meet another person at random, it's almost certain that he or she will know someone who knows someone whom you know.

But, as Paulos noted in his book, whether any two random people will share with one another their lists of everyone they know and discover their common linkages, "is another, more dubious matter."

Actually, the small-world phenomenon was once put to a real-world test by a psychologist named Stanley Milgram. That experiment has become a classic; Paulos, Gardner and others cite it, based on Milgram's account in the May 1967 issue of *Psychology Today*.

Milgram began with a random group of people. Each one was given a document to transmit to someone in a distant state whom the starting person did not know. He or she was to mail the document to a friend who was most likely to know the target person; that friend was to send it to another, and so on until the document reached someone who personally knew the target person.

How many people will handle such a document before it reaches the target person? Before he actually conducted the experiment, Milgram took a poll and found that most people guessed that about 100 people would be involved before the document actually reached the intended person.

In fact, Milgram found that the number of intermediate links ranged from two to 10, and the median number was five.

Mathematician Michael Guillen wrote, in the October 1983 issue of *Psychology Today*, that when he has a small-world experience, he thinks of it in terms of winning a lottery. A seemingly unlikely encounter, he said, is no less weird or mystical than having a winning lottery ticket.

But people are amazed, and sometimes see something more fateful than mere coincidence, when a small-world experience occurs to them.

Throughout the nation, Guillen said, strangers are continually crossing paths and thinking nothing of it because the noteworthy element of coincidence is missing.

There's a good chance, however, that hidden linkages lurk in many such encounters.

For, after all, it really is a small world, statistically speaking.

When Your Luck Runs Bad, Blame It on Probability
AUGUST 9, 1987

We all occasionally have one of those days.

Such was the case some years ago when a newspaper photographer was on an assignment on the Medical College of Virginia West Hospital's eighth floor.

After he completed the assignment, he took an elevator to the first floor and started walking to his car when he remembered that he left his hat in the room where he had just taken pictures, eight floors up.

He returned to the eighth floor, retrieved his hat, caught the elevator back to the first floor and walked out to his car—which had a parking ticket on it.

When he returned to the newspaper's photo department, he discovered that he had not put film in the camera that he used on the assignment.

Some might call it a run of bad luck.

That's what some say has been happening with Delta Air Lines in recent weeks. The airline has been plagued by a series of seemingly unconnected events, such as one of its planes landing at the wrong airport, another landing on a wrong runway, an engine shut-down during a flight, and reports of near-collisions.

A recent Associated Press report on Delta's problems quoted a wide range of people, from Delta's own spokesmen to officials from the

National Transportation Safety Association, a consumer group, as saying that there appears to be no underlying pattern to the problems.

"An incredible streak of bad luck" was how one Air Line Pilots Association spokesman summed up Delta's plight.

But what, exactly, is "bad luck"? In some cultures, it may be viewed as the result of a curse, or a just reward for some past action, or the result of violating a superstition, like breaking a mirror.

A more rational view is the one that's rooted in the mathematical discipline of probability.

Essentially, any series of individual events is likely to contain runs of particular outcomes. Flip a coin a few dozen times, and you're likely to have runs of four or five heads (or tails) in a row, even though the overall result of the tossing is around 50 percent heads and 50 percent tails.

Perhaps a mathematically inclined person could put the Delta streak into some sort of perspective, taking into account the numbers of pilots in the airline, the numbers of flights, the numbers of routes, runways and airports, the numbers of mechanical and electrical systems on each airliner and the numbers of rules and regulations that apply to commercial flights.

The numbers of possibly interacting events and systems could easily run into the thousands daily, and even those not mathematically inclined can have an intuitive grasp of how unfortunate outcomes can sometimes pile up.

The Air Line Pilots Association spokesman was quoted as saying that out of the thousands of daily operations, occasional goofs do occur. "Normally," he said, "something might happen to say, United one day and Pan Am the next and maybe to Northwest the next. It is all spread around and nobody takes any notice."

It's a rare day that such an event doesn't occur, according to the National Transportation Safety Association official. "Inexplicably," he said, " they are all happening to one major airline in a short period of time." He went on to say that Delta has about 6,500 pilots and that only about four of the recent goofs were clearly the fault of Delta pilots.

"That's four or five captains out of a pilot pool of 6,500. That isn't bad," he said.

Even if Delta's problems are not related events, people tend to tie them together anyway. The news coverage that's given to each latest incident promotes the impression that all of the events are somehow related.

The same sort of thing happens in the sports world. There are perceived "hot streaks" and "cold streaks" for players in certain sports, like basketball. A player may be considered to have a hot streak when he or she makes three or four consecutive shots; fans believe the player's on a roll, sportscasters may comment on it, and fellow players may try to feed the ball to the "hot" player.

But is a player really more likely to hit the basket after a previous hit? Some mathematicians studied the statistics of four dozen 76ers games during the 1980-81 season and found that the team's members were actually more likely to hit the basket after a miss than after a previous hit. For one player, Darryl Dawkins, the mathematicians found he made about 70 percent of the shots he took after misses, and 57 percent of those he took after hits.

The mathematicians' conclusion: The "hot hand" doesn't really exist. In any string of events, there will be runs of hits or runs of misses, just like there will be runs of heads or tails in a series of coin tosses. But, they found, the numbers of hot streaks and the numbers of cold streaks were just about what chance alone would dictate.

Players tend, in the long run, to perform at a level that's usual for them. And that may be true for airlines and individuals. We may have our good days and our bad days, and if we don't let the bad days shake us and the good days lull us, we probably will turn in performances that hover around a particular average that's about right for us.

Did O's Break Law of Averages?
May 1, 1988

More than 40 years ago, Robert M. Coates wrote a delightfully imaginative short story about what could happen when the law of averages breaks down. His story was titled "The Law," and it was published originally in the Nov. 29, 1947, issue of *The New Yorker*.

The "law of averages" is a popular expression for the tendency of a mass of erratic events to follow more or less regular patterns.

For example, traffic engineers can't predict the routes of individual cars, but they do know roughly how many cars will use a particular street or cross a given bridge every day. Restaurant owners know roughly how much each menu item will be called for each day, although they are unable to predict what particular individuals will order. Store operators know approximately how much of each item should be stocked, based on average demands.

But, according to Coates' story, when the law of averages began breaking down, hordes of motorists used a particular New York bridge one night; dinner customers all ordered the same thing, and one store owner noted that over a four-day period, 274 successive customers asked for a spool of pink thread.

In the contemporary sports world, the question may be raised whether the law of averages broke down for the Baltimore Orioles, which had a noticeably long losing streak going back to the beginning of the baseball season. When the team finally won Friday night, the streak ended at 21.

One may wonder whether indeed the law of averages has its lapses in such cases. And closely related to that issue is the popular belief that as the losses pile up, so do the odds that the team will win the next game, or at least the one after that.

Behind that belief is the vague assumption that there's some sort of superintendent of luck somewhere who keeps tabs of wins and losses, and when an uncommonly long string of losses (or wins) occurs, the luck superintendent makes an immediate correction to keep the law of averages honest.

People who specialize in the mathematics of statistics and probability emphatically state that this is wrong thinking. Truly random events have no memory, they point out, of what happened previously; one event is not influenced by what happened in the past.

Flip a coin, for example. If it's really a random toss and if the coin is not weighted in any way, there's a 50 percent chance it will land heads up and an equal chance it will land tails up.

What the so-called law of averages claims is that the more you toss the coin, the closer you will get to a 50-50 distribution of heads and tails. If you flip a coin 10 times, you may get seven heads and three tails, for a 70-30 break.

If you flip it 100 times, you may get 54 heads and 46 tails, in which case the distribution will be 54-46. Flip it 1,000 times and the results may be 502 heads and 498 tails, for a 50.2-49.8 distribution.

Within those 1,000 flips, however, may be runs of tails or heads. A dozen tails, for example, may occur in a row. (The chances of that happening are one in about 4,100 tosses.) But that doesn't mean that the chances of a head will be increased for the next toss.

In his book, *Lady Luck, The Theory of Probability*, mathematician Warren Weaver said he once asked his wife what she would expect to happen if a coin turned up heads eight times in a row. He quoted her as saying, "It just stands to reason that you are much more liable to get a tail on the ninth toss—for everyone knows that in the long run the heads and tails balance out, and some balancing is obviously overdue."

Even though it sounds plausible, and even though many people would agree with her, Dr. Weaver noted that the chance of getting heads on the ninth toss is 50 percent—exactly what it was on the first, second, third and every toss. What happened before has no bearing on the next toss.

Similar "plausible" reasoning has been entering recent small talk about the Baltimore Orioles—namely, the longer a losing streak, the greater will be the pressure on the law of averages to give the losing team a win.

But, as Dr. Weaver and others would point out, the law of averages is not a compulsory force like gravity.

In fact, one may even question how much the law of averages can be invoked to describe long losing streaks of a baseball team. Each coin toss is independent of the previous one, but in a baseball game, previous losses can psychologically affect the members of a team.

A baseball game is not a pure event of chance like a coin toss. A baseball game (or any other game of sports) involves perceptions, strategy, moods, skill, confidence and many other intangibles.

So there's more to Baltimore's record than the law of averages.

Probably.

In The Lottery, Either You Win or You Lose
SEPTEMBER 6, 1993

In one of his books, mathematician John Allen Paulos tells of the barber whose theory of lottery probability goes something like this: Either he will win or he won't, so his chances of hitting the jackpot are 50-50.

That's one thing about lotteries. They evoke many interesting hypotheses about chance, luck and numbers.

And occasionally, numbers that someone considers to be personally lucky are winners, thus reinforcing the notion—and hope—that luck is not just a matter of chance.

Such was the case recently with the winner of Virginia's $24.5 million Lotto jackpot. Gerald L. Ranes and his wife, Rita Sue, of Newport News selected six personally significant numbers that matched the six that were drawn randomly from 44 numbered balls.

The winning combination was 1-8-10-28-39-42. The barber's logic notwithstanding, the chance that a given combination of numbers wins in the Lotto game is 1 in 7.1 million.

On the Raneses' ticket, 1, 8 and 39 represent his birth date; 10 and 28 represent the date they met, and 42, they said, has a significance they will not reveal publicly. (It does call to mind the enigmatic answer that the computer, Deep Thought, gave to the great question of life, the universe and everything in Douglas Adams' book, *The Hitchhiker's Guide* to *the Galaxy*. The answer, the computer said without elaboration, is "forty-two.")

Actually, Virginia Lottery records show that more than half of the winning tickets in Lotto, which began in early 1990, have contained impersonal numbers—numbers that were chosen by electronic devices in Lotto ticket machines around the state.

Those devices are computer chips that generate numbers randomly. If players want the machine to select six numbers, they mark a box labeled "easy pick" on the ticket slip and the machine's random-number generating chip goes blindly to work. The player can mix easy picks and personally selected numbers on the same ticket.

Paula Otto, the lottery's spokeswoman, said last week that 69 of the 128 winning combinations to date were selected by the easy pick computer chips. The remaining 59 were number combinations that the winners picked themselves, but not all of those winners used numbers that had any personal significance to them.

For instance, David Snyder of Lynchburg, who won nearly $11 million in September 1990, used his personal computer and a commercial software program to generate his winning numbers, Otto said.

But she said Maurice and Esther Coats of Virginia Beach used a considerably less sophisticated type of random-number generator to select the numbers for their winning ticket in November 1991. They wrote the numbers 1 through 44 on slips of paper, put them into a coffee can, shook the can and drew six numbers. And they won $2.5 million.

Robert and Jane McGaw of Gloucester, meanwhile, became personally attached to an impersonal combination the computer gave them the first time they played, back when Lotto began, and they stuck with that combination from then until October 1990. Their numbers came up, and the McGaws won $13 million. At the time, McGaw said of the numbers the computer first gave him, "I guess I'm superstitious. I figured if I ignored that combination, it would come up."

Then there were the personal combinations that worked—sometimes with updated changes and sometimes without.

Lotto's first winner, Anthony Palermo of Virginia Beach, had been playing the lottery in other states for years, using days of the month of his own, his wife's and his two daughters' birthdays; a 2—because they have two daughters; and the number of years the Palermos were married.

Otto said that last number changed from a 9 to a 10 just before he bought his winning Lotto ticket. He changed it on the ticket and won $7 million.

But when Tina Saunders of Roanoke County turned 26, her husband refused to change the 25 in the Lotto ticket he bought for her the following week. The numbers she selected included her age, the ages of her siblings and the years her parents were born.

Even though her age changed a week earlier, Tina's husband, Dwayne, told her not to change the number on her ticket. The result: a $1.7 million Lotto win.

If there's a message in all of those anecdotes, it is simply that chance can conjure up some interesting and ironic twists when tens of thousands of people are involved in something like selecting numbers for a game of chance.

Another message is that even lucky numbers win sometimes.

Full Moon and Wacky Behavior
MAY 19, 1991

News people, policemen, hospital workers and others who deal directly with lots of people swear there are tides in human behavior that, like the Earth's watery tides, are affected by the moon.

"Must be a full moon," they are likely to say when there seems to be a higher-than-usual number of crazy calls, violence, accidents or other kinds of odd behavior.

It's an old notion. The alleged connection of the moon with erratic or wild behavior is embodied in such words as "lunacy" and "lunatic," which are based on the Latin word for moon, luna.

In the 1700s, the English judge and author William Blackstone took note of the moon connection. "A lunatic," he wrote, "is one who hath . . . lost the use of his reason and who had lucid intervals, sometimes enjoying his senses and sometimes not, and that frequently depending upon the changes of the moon."

And, of course, there are the folk tales about the full-moon vampires and werewolves that continue to provide themes for horror and comedy-horror movies, such as *I Was a Teen-age Werewolf* and *An American Werewolf in London.*

According to at least one astrologer, and to many laymen, it makes sense that the moon should affect people's behavior since it has a powerful effect on the Earth's oceans. Some who make such grand leaps of logic also throw in the irrelevant fact that the human body, like the planet Earth, is about 80 percent liquid and 20 percent solid.

Despite the anecdotes, the folklore, the superstitions and the curious logic, however, there doesn't seem to be any good data that show a link between the full moon and increased craziness among the population.

Calls to crisis hot lines have been plotted against phases of the moon. So have murders in several major American cities. So have admissions of patients to psychiatric wards or hospitals. So has the incidence of various crimes, including rape, assault, robbery, disorderly conduct, drunkenness and family rows.

And none of the scientifically and statistically rigorous studies shows that the full moon is associated with any such behaviors. One researcher, in fact, titled his study, "The Moon Is Acquitted of Murder in Cleveland," according to a past issue of the *Mayo Clinic Health Letter.*

A particularly good analysis of the full-moon-and-craziness belief is included in a book that takes a scientific look at astrology, *The Gemini Syndrome*, co-authored by a University of Virginia astronomer, Dr. P.A. Ianna. The other author is Dr. R.B. Culver of Colorado State University.

They cited one of the relatively few studies that does claim to find a relation between homicides and the full moon, but they also said there were weaknesses in the study. For one, Drs. Ianna and Culver said, the researchers didn't take into account factors other than the full moon that could account for periodic outbursts of odd or violent behavior.

Take weekends, for example. People get paid. They're off. And they tend to mix, drink, party, get into trouble, be on the road, have accidents. A check with the Charlottesville-Albemarle rescue squad, Dr. Ianna and Dr. Culver said, showed that the numbers of emergency calls do indeed go up during weekends.

And during an average year, one out of four full moons occurs on a Friday or Saturday night.

The heart of the full-moon belief seems to be the fact that the moon pulls on the Earth and its inhabitants. The relevant principle here is Newton's law of gravitation, which says that every object in the universe tugs at every other object. The strength of that pulling force diminishes with distance—with the square of the distance, in fact.

Using Newton's equation, scientists can calculate how much gravitational force the moon exerts on an individual person, as well as on oceans. The pull on an individual works out to a few one-hundredths of an ounce, which is roughly the gravitational force that a medium-sized (say five-story) office, hospital or department store building exerts on passers-by.

The major reasons the moon causes tides include the facts that its gravitational pull is constant, and that the vast amounts of liquid on the moonward side "feel" and respond to the force more than the oceans on the opposite side of the Earth.

It is true, Drs. Ianna and Culver noted, that a number of sea creatures do respond to the moon's phases. Some have gravity-sensing organs that are sensitive to tiny changes in gravitational forces; some apparently respond to the increased light of the full moon. And there are some apparently moon-linked behavior patterns among some creatures that scientists don't understand yet.

But do people act crazier during nights of the full moon because the moon is full? Most scientists who've studied the question think not.

As the *Mayo Clinic Health Letter* noted a few years ago in its review of the full moon's alleged effects on people: "If the moon does hold power, it is mostly the power of suggestion."

MATH—THE TRAP

Mathematics is a trap. If you are once caught in this trap you hardly ever get out again to find your way back to the original state of mind in which you were before you began to investigate mathematics.
E. COLERUS, *From Simple Numbers to the Calculus*

There's a strange, fascinating connection between mathematics and the real world—mathematics is an abstract product of the mind, but scientists often find it useful in describing the external world. That intriguing power has been one of the abiding allurements of mathematics for me over the years, and so has been its ability to stimulate the imagination. What does a hypercube look like? How is it that the arithmetic of infinity can lead to the paradoxical conclusion that infinity plus infinity equals infinity? And how is it that straight pins and straight lines, plus chance, can lead to the value of pi, the fundamental ratio of the circle?

Nothing: Little Things Apparently Mean a Lot
SEPTEMBER 27, 1987

There has been a lot of talk about nothing recently.

It has become topical because it figures into some current theories about how the universe may have come into being. According to those theories, our universe—galaxies, stars, planets, civilizations, people, computers, video games and Slurpies—ultimately arose from what we laymen would consider a vacuum, or nothing.

But, according to the physicists who study our universe's beginnings, what we might consider a vacuum is actually a foamy, seething stratum of reality that can't actually be compared with anything in our experience.

It is a quantum vacuum, seething with transient activity, including particles that arise and vanish randomly. It has nothing, or nothing to which our imaginations can relate.

It was in that strange cauldron of unimaginable activity, many cosmologists believe, that a random fluctuation occurred and, within an instant, became a runaway event that quickly gave rise to our embryonic universe.

From such theories come the staggering notion that at least once in the history of the universe, something did come from nothing. Indeed, as noted by Paul Davies, a British physicist, in his 1984 book, *Superforce*: "Today we can argue that everything has come out of nothing. Nobody needs to pay for the universe. It is the ultimate free lunch."

This is not the first time that nothing has become a matter of interest.

Sometime between 1,000 and 1,300 years ago, some unknown mathematicians in India introduced the notion of nothing into the numbering system; that notion was the zero.

The idea of a zero in the numbering system is not obvious. Apparently, it did not occur to anyone earlier, or if it did, the idea did not endure for long. The Greeks, the Romans and other past civilizations that had attained a great deal of sophistication in mathematics and engineering did not have a zero in their numbering systems, and they seemed to get along fine.

Apparently, the idea of nothing in math was not a compelling one because there was not a compelling need for it in everyday life. After all, noted one mathematician and philosopher, Alfred North Whitehead, no one goes out to buy zero fish.

The number zero, according to Whitehead, "is a very subtle idea, not at all obvious." To him, zero is a number, no different from 1, 2, 3, or any other.

So what is the value of zero in math?

One use is to indicate that nothing's there.

In our arithmetic system, the value of a number is indicated by its place. In the number 5,281, for example, the 5 is in the thousands place, the 2 is in the hundreds place, the 8 is in the tens place and the 1 is in the units place. So the number 5,281 really means there are five one-thousands, two one-hundreds, eight tens and one unit.

But suppose there were no one-hundreds in that number. We would write it 5,081, with the zero indicating that nothing is in the hundreds place.

In ages past, according to students of math history, cultures that used the place-notation system would leave a space between the 5 and the 8 in that example to indicate that there were no hundreds. But that could easily lead to confusion and mistakes in reading the values of numbers, given the wide range of handwriting styles.

Also, how could one know whether 55 was fifty-five, or 550 or 5500, or whether 5 was five, or 50 or 500? The zero, denoting empty places or places with nothing in them, made the expression of numbers clear.

Whitehead said the people who introduced zero for that purpose probably had no mental conception of zero; they most likely introduced it merely as a symbol to communicate the message that there was nothing in the place where it stood. Only gradually, he said, did the notion of zero as a number sink in.

Sometime around the 16th century, the zero began contributing to the growth of algebra when mathematicians began setting sums and differences of X's and Y's equal to zero. That may appear to be a subtle, almost trivial step, according to Whitehead, but it was an important and powerful one in the development of mathematics.

He and others point out that the zero is essential to modern mathematics, making possible investigations that could not have been conducted without it.

Therefore, so much has been contributed to our lives by nothing.

In fact, our very existence—if the physicists are correct in their theories about the origin of the universe—is due to nothing.

In which case, there may be a deeper meaning to the phrase: "Thanks for nothing."

Are There Degrees of Infinity?
NOVEMBER 29, 1970

A graduate student at Trinity
Computed the square of infinity.
But it gave him the fidgets
To put down all the digits,
So he dropped math and took up divinity.
— Anonymous

In a world where we are constantly being made aware of beginnings and endings in everything the senses perceive, it seems odd that a concept of infinity should enter human minds and become an important element of human thinking.

But it has. And interestingly enough, it appears that men first grap-

pled with the concept of boundlessness centuries before the idea of noth-
ingness—zero—entered their number schemes.

The introduction of zero into the number system was, of course, a
great stride and a powerful contribution, according to mathematicians.
But zero has probably never had the inscrutable lure that infinity has had,
particularly in the minds of people other than mathematicians.

The concept of infinity, its history shows, has been such an impen-
etrable one that the sharpest thoughts of great people through the ages
have glanced off it, ricocheting into a variety of disciplines, from religion
to philosophy, from geometry to cosmology.

And infinity, symbolized by a horizontal figure eight (∞), has turned
out to be more than an elusive abstraction. Rather, it is relevant, because
lurking somewhere, directly or indirectly, within the technical accomplish-
ments of modern humans are arithmetical processes that involve infinity.

By and large, whenever arithmetic is applied to such fields as
physics, mechanics, geometry and statistics, "infinite processes," as the
mathematicians call them, are somewhere, somehow involved.

The concept of infinity is a concept of endlessness, such as is
involved with the process of counting. One can start counting, one, two,
three, four... and never reach a last number. Wherever a person stops
counting, one can be added to that number to give yet another number.
And so on.

But the matter of infinity doesn't stop there, although it did more or
less stall there until the latter part of the last century, when a German
mathematician, Georg Canter, began charting the puzzling and absurd
world of infinity.

He showed a way of "counting" infinity, and he found degrees of
infinity—some infinities are greater than others.

He based his ideas on the seemingly simple process of counting,
the foundation for arithmetical operations. When a person counts something, say the eggs in a carton, he or she matches a number with each
egg in sequence.

Or, the person may immediately count the eggs if he or she knows
that a carton has depressions for each one and that there are 12 such depressions in each carton. If each depression is filled, then there are 12 eggs.
There is a one-to-one correspondence between depressions and eggs.

And the person can quickly compare the numbers of eggs in two
cartons. If the depressions in both cartons are filled with eggs, then it is

known that the number of eggs in one carton is equal to the number of eggs in the other.

In a similar way, different kinds of infinities can be compared, using the numbers, one, two, three, four...to match up with the elements of a collection of numbers.

The sequence, one, two, three, four... is an infinite collection. How about the sequence of even numbers, two, four, six, eight....?

This sequence can be matched, element for element, with the one, two, three... sequence; hence, the collection of all even numbers is infinite. By similar reasoning, so is the collection of odd numbers, and so is the collection of squared numbers.

More generally, if a set of numbers can be matched one by one, with the natural numbers, one, two, three... then that set, like the natural numbers, is infinite.

And therein lies one of the many strange things about infinity. The even numbers (or odd numbers, or the squares of numbers) represent a part of the total collection of natural numbers. But the even (or odd, or square) numbers are infinite.

Thus, in the case of infinity, a part is equal to the whole. Infinity (even numbers) equals infinity (all natural numbers).

Further, if all the even (or odd, etc.) numbers were subtracted from the collection of natural numbers, the remainder would still be infinity. Infinity minus infinity equals infinity.

Similarly, if the infinity of fractions between the whole numbers are added to the collection of natural numbers, the sum is infinity. Infinity plus infinity equals infinity. Similar apparent paradoxes also appear with multiplications.

Such peculiar properties of the world of infinity have been anticipated by some of the world's mystical religions, thus giving a measure of credibility to the limerick quoted above.

Tobias Dantzig notes in his book, *Number—The Language of Science*," a reference in Sanskrit writings to "...the Infinite, Invariable God who suffers no change when old worlds are destroyed or new ones created, when innumerable species of creatures are born or as many perish."

The Conceivable but Unimaginable Hypercube
AUGUST 21, 1977

Sometime, when you have nothing else in particular to think about, you might try to imagine what a tesseract looks like.

A tesseract is a four-dimensional version of a cube. It is what a cube would be in hyperspace.

And hyperspace has been a device for many a science fiction story, including the popular movie, *Star Wars*. It is hyperspace, for example, that the heroes enter in the nick of time—amid a brief enlargement of a rapidly passing star field—as they are being chased by the space fighters of the bad guys.

Actually, the idea of hyperspace has been around for at least a century. It was born in the field of geometry, when mathematicians began conceiving of geometries of dimensions higher than the three that apply in our world of experience—length, height and breadth.

Although mathematicians can conjure up geometries and equations that would apply to four, five, six, seven and more dimensions, it is the fourth dimension that has particularly caught the fancy of various people in the past.

Interest in the fourth dimension appeared to be especially high around the turn of the century. H.G. Wells wrote a story, "The Plattner Story," that invoked the fourth dimension.

Scientific American, in 1909, offered a $500 prize for the best popular explanation of the fourth dimension, with 245 essays on the subject submitted. Some of those essays are in a book, *The Fourth Dimension Simply Explained*, that has been republished by Dover Publications, Inc.

Spiritualists of that era were claiming the fourth dimension as the abode of disembodied spirits. It's no wonder that the fourth dimension has stimulated widespread interest. After all, amazing and magical events can occur there. Take a lefthanded glove, for example, and flip it into the fourth dimension. It will vanish, then reappear as a righthanded glove.

If you could somehow exist in the fourth dimension, you would be able to come and go from locked, closed rooms without passing through the walls, doors or windows. You would be able to remove the contents of an egg without breaking its shell.

Imaginative explorations of hyperspace are usually explained by analogy. Imagine a world of creatures that are perfectly flat and whose world is perfectly flat. They are two-dimensional creatures; they know length and width, but there is no height in their world. Height for them would be a mysterious third dimension that they cannot sense or know.

The flat world creatures can know of the various plane figures. They can know of circles but not spheres; they can know of triangles but not pyramids; they can know of squares but not cubes.

They can also know of handedness. They could have, for instance, an L-shaped figure where the bottom arm points to the left, and they could have one in which the bottom arm points to the right. But there is no way in which the flat world creatures could make the two L figures superimposable, that is, both pointing the same way.

But we from the world of three dimensions could reach into their two-dimensional world, pick up one of the L figures—bring it into the third dimension—turn it around and return it, pointing the same direction as the one that remained in the flat world.

For the flat creatures, something magical happened; the figure vanished from their world, then reappeared in reversed form. It is from such an analogy that various writers speak of turning lefthanded gloves into right handed ones merely by taking the lefthanded one into the fourth dimension, flipping it around and returning it as a right handed one.

And it is by such an analogy that various writers describe such things as escapes from closed boxes, rooms or eggshells via the fourth dimension. For flat world creatures, a complete circle or square or triangle would be a secure enclosure. But for us in the three-dimensional world, it's no problem whatsoever to pluck something out of their enclosures. We merely lift it straight up, into the third dimension.

In the same way, escape from or access to any of our three-dimensional enclosures would be easy and nondestructive via the fourth dimension. A hyperspace doctor, the analogy suggests, would be ideal. He or she would be able to inspect our heart, liver and other internal organs without opening us up.

The question of "where" the fourth dimension exists is best answered by saying it exists only in the playful musings of geometricians. But the human imagination—even that of the geometricians—falls short of visualizing what even the simplest four-dimensional object would look like.

And that brings us back to the tesseract, the hypercube. One can build up to it like this: Take a straight line and move it perpendicular to itself. It will trace out a two-dimensional square. Move the square perpendicular to itself. The result will be a three-dimensional cube.

Now take the cube and move it perpendicular to itself. The three-dimension conditioned imagination fails at this step, but the result would be a hypercube. It probably won't help, but geometricians say the tesseract has 16 corners, 32 edges, 24 square faces and is composed of eight three-dimensional cubes.

The geometricians also can tell you what its three-dimensional "shadow" looks like, and they can show you how the hypercube appears when it is unfolded in our world. (There's a Robert Heinlein science fiction story , "And He Built a Crooked House", about a man who built his home in the shape of an unfolded hypercube. Unfortunately, an earthquake caused it to fold up.)

But despite the geometricians' descriptions, virtually no one has claimed to have clearly visualized a tesseract in the mind's eye, no more than a hypothetical two-dimensional creature would be able to visualize a cube.

It's a curious example of the conceivable being inconceivable.

Almost Anything Can Be Reduced to Information
JANUARY 24, 1994

The information superhighway that everyone's talking about probably started with those early people who paused to draw pictures on their cave walls. One wonders where the road will lead in the other direction, the future.

Will humans themselves be someday reduced to bits of information, storable in data banks or capable of being transmitted at the speed of light, like faxes?

Science fiction writers, who sometimes have served as guides for the future, have already considered such fantastic notions. It all triggers random thoughts about the nature of information itself.

A simple view is that information is anything that can communicate something.

Your presence at a particular function or place could be a piece of information. So could your absence.

A raised eyebrow can communicate.

Rolling the eyes upward can be a silent comment. So can knitted eyebrows. So can a smile, a frown, a reddening face.

Puffs of smoke can convey information. So can dots and dashes. So can grooves in a phonograph record, holes in punch cards, patterns in magnetic tape.

Black means one thing. White means another.

Red means stop; green means go and yellow means caution (or, as noted by the alien in the movie, *Starman*, yellow often means "speed up" to Earthbound drivers.)

Information can also be conveyed by a yes or a no, a nod or a shake of the head. It can be communicated by a one or a zero, an on-switch and an off-switch.

That's behind the magic of computers, which translate information as varied as gory video games to words from languages of off and on signals. As the ubiquitous computer demonstrates, practically any kind of information can be converted into digital information, or information composed of the simple digits of ones and zeros.

Previous ages have been identified by their prime material or activity—the Iron Age, the Bronze Age, the Industrial Revolution, the Atomic Age.

Now, it's the Information Age.

Even biology and medicine are into it. What is DNA but a chemical code that contains information on how to make an entire organism, such as a human? What's a gene but a piece of information from DNA on how to make a specific protein or enzyme?

Scientists are now turning to treatments based on correcting faulty information, by inserting normal genes in patients to make up for problems caused by faulty ones. Gene therapy is still in its early stages, but scientists say they are encouraged.

Also, one of the major "big science" projects now underway is the genome project, in which all human genes are being charted. The final product, some years away, could be an informational map of a human being.

Ultimately, practically everything comes down to information, including ourselves.

Could we—or the information that defines us—be someday transported around the planet or solar system or the universe?

Maybe so, suggests Arthur C. Clarke, the writer who is credited with coming up with the idea of communications satellites back in the 1940s. Right now, the barriers seem formidable, he noted in an essay, "World Without Distance," in the 1984 edition of his book, *Profiles of the Future.*

Clarke said that although he has no idea how it can be done, he believes that it will be possible one day for people to move "from Pole to Pole within the throb of a single heartbeat."

But, as he pointed out, it may not be necessary to be transported. If information about all the sights, sounds, smells and feel of a place can be transmitted, why would we need to be transported there in the first place?

Information, after all, can go both ways on the information highway.

Jefferson's Pendulum Could Have Given Us Metrics
MAY 24, 1993

It's been 250 years since his birth, and Thomas Jefferson's name is still bandied about in science journals.

Jefferson, as a visit to Monticello or reading almost any book on his life makes clear, was as much a scientist as he was a statesman, architect and philosopher.

He invented scientific and mathematical instruments; he kept detailed weather records; he became well versed in Virginia's plant and animal life, and he took such a scholarly interest in fossils that he has been called the father of American vertebrate paleontology.

"Science is my passion; politics my duty," Jefferson once wrote. And although most of the ado that's being made over Jefferson in this 250th anniversary of his birth is over his political career, at least some of the fruits of his passion for science are also remembered.

His name popped up recently, for example, in the American Institute of Physics' publication *Physics Today* in a discussion among some letter writers about the origin of the meter, the metric system's fundamental unit of length.

When Jefferson was Secretary of State in George Washington's administration, noted Ted Uzzle of Oklahoma City, Congress asked his office to develop a weights and measures system for the new country.

Jefferson proposed that a pendulum be the basis of a measurement system, Uzzle's letter recalled. Silvio A. Bedini explained in his book,*Thomas Jefferson: Statesman of Science*, that Jefferson believed a fundamental length ought to be based on a natural phenomenon that could be readily reproduced anywhere.

A pendulum fits that bill because its length is related to the duration of its swing. The longer the pendulum's arm, the more time it takes to swing from one side to the other. Length of the arm and the pull of gravity are the main things that affect the timing of a pendulum's back-and-forth arcs.

Jefferson proposed that a measurement system be based on the length of a pendulum that takes one second to make one swing, or two seconds to return to its starting point. The length of the pendulum's arm in that case is 39.2 inches.

Such an ideal pendulum could then be related to what physicists call a physical pendulum or a compound pendulum—a real-world object like a rod that also makes one-second swings when it's pivoted at one end.

Jefferson's proposal makes use of both types—the ideal and the physical pendulums.

He proposed that the physical pendulum be a metal rod. There is a point in a rod or other such objects where all of its mass is seemingly concentrated. It's called the center of oscillation; it's the point on a baseball bat where a batter aims to hit the baseball so no stinging vibrations are transmitted to his hands.

If the distance of the rod's pivot point to the center of oscillation is 39.2 inches, the rod will make one-second swings from one side to the other. It is, in other words, equivalent to the simple pendulum of that length.

But the rod is actually much longer; the math and physics of it show that 39.2 inches is only two-thirds of its total length. The rod in that case turns out to be 58.7 inches long.

Jefferson said the standard of length, then, should be twice the length of a rod that makes one-second swings or three times the length of a pendulum that makes one-second swings.

In his second report to Congress, made in the spring of 1790, Jefferson proposed that the 58.7-inch rod be divided into five equal units called feet. The new foot would be about one-quarter inch shorter than the conventional foot. Each new foot, in turn, would be divided into 10 equal parts, or new inches that would be slightly longer than our present inch.

Congress rejected Jefferson's proposals. France, meanwhile, developed its own system about the same time and defined the meter—the system's fundamental unit of length—as one ten-millionth of the distance from the North or South Pole to the equator, along a meridian that runs through France.

The length of the meter is 39.4 inches (now defined as the distance light travels in a vacuum in a specified fraction of a second).

And guess what? That's amazingly close to the 39.2-inch length of a pendulum that ticks off a second per swing.

As one *Physics Today* letter writer noted, the pendulum idea suggested by Jefferson and some other scientists of the day would have been a lot cheaper than paying for an expedition to survey a meridian that runs through France.

Case of Pi and Pins Makes Point for Labs
JUNE 8, 1995

Of all the chemistry and physics labs I had during college, several made a lasting impression, especially the one in which I discovered pi with straight pins.

Many of my science course labs were not so memorable. In fact, I can recall little specific about them, except that they posed some troublesome discrepancies between what was supposed to happen and what actually happened.

Labs are where theory meets the real world, and the real world often wins. Things don't always work according to the neat equations in a textbook because the real world is more complicated than the equations let on.

And that's a necessary lesson for prospective scientists and non-scientists alike to learn, as pointed out by a resolution that the Virginia Academy of Science recently adopted about the importance of labs in science courses. The academy adopted the resolution in light of a gradual disappearance of the lab in higher science education.

If certain physics labs had been discontinued in my time, I would have missed some experiences that have stuck with me for several decades. One was the straight pin experiment that was the introductory assignment for an upper level physics lab course.

Each of us was given a bunch of long pins—20, as I recall. We measured each one with calipers and figured an average length, then on a poster board we drew a series of parallel lines. The space between the lines was equal to the average length of the pins.

The experiment involved tossing the pins onto the poster board and counting the number of pins that crossed a line on each toss.

There were only five of us in the class, and we worked on the lab floor. We laid our poster boards on the floor and we knelt, tossing our pins on them as if we were back-alley dice tossers. We tossed, we counted and we tossed again.

Our instructor told us that the more random the toss, the better our data would be. So we soon developed flourishes in our tossing of the pins. We threw them in the air; we dropped them from standing positions and we made various kinds of paper cups for shaking and throwing.

Here was the intriguing point of that experiment: We could derive pi, the fundamental constant of the circle, from our pin tosses. Straight pins and straight lines magically yielded the most famous ratio in the world, 3.1416, the ratio of the circumference of a circle to its diameter.

The reason is that the probability of a pin crossing a line is 2 divided by pi. The proof of that requires considering all of the possible positions a pin can take between any two lines that are spaced a pin's length apart.

(For a good discussion of how it works, look in a book called *One Two Three . . . Infinity* by the late George Gamow. He talks about matches instead of pins, and instead of a poster board with parallel lines, he talks about the matches intersecting stripes on an American flag.)

It turns out we were rediscovering a finding originally made by the 18th-century French naturalist Compte de Buffon. And, we were following the efforts of others. At the beginning of the 20th century, an Italian mathematician tossed a needle onto a parallel line grid more than 3,400 times and reported getting a value of pi of 3.1415929. He was off only in the seventh decimal place, where the 9 should be a 6.

I don't recall how well I did, nor do I recall how many times I cast the pins. But I do remember being fascinated that pi can come from such an unlikely experiment.

I could have read about it in a book.

But I doubt I would have remembered it so vividly decades later if I hadn't actually knelt on the floor of a physics lab and conjured up pi with random tosses of pins on a poster board.

Fermi Mastered Art of Making Educated Guess
SEPTEMBER 28, 1995

Parade columnist Marilyn vos Savant recently posed the following puzzle: If you wanted to camouflage yourself for an undisturbed nap by painting your body the same color as your sofa, what square footage must you cover?

She gave the answer, but she didn't show her work.

Such a problem calls for a ballpark estimate, sometimes called an educated guess. One approach is to consider the body to be a rectangular box and to make some rough assumptions about its average dimensions, such as 69 inches high, 14 inches wide and 7 inches thick for a man, and 64 inches tall, 12 inches wide and 6 inches thick for a woman.

Based on those assumptions, a man has about 21 square feet of surface area and a woman, about 17 square feet. Vos Savant's answers were about 20 square feet of skin for a man and 17 for a woman.

The art of the educated guess was taken to legendary heights by the physicist Enrico Fermi, and many problems involving ballpark estimates are sometimes called Fermi problems.

In his book, *The Fermi Solution*, Dr. Hans von Baeyer of the College of William and Mary recounts the well-documented story of Fermi tossing torn bits of paper into the air when the first atomic bomb exploded in a New Mexico desert in 1945. Fermi noted how far the bomb's shock wave blew the paper shreds and calculated, in his head, the approximate energy released in the blast.

Fermi's estimate was confirmed several weeks later when data from various instruments around the blast site were analyzed.

Von Baeyer said a Fermi problem has a characteristic profile. When you first hear it, you don't think you have enough information to solve it. But it can be broken down into separate problems that can be solved based on what you already know, and the resulting estimate is close to the real answer.

For example, based on what you already know, what is the circumference of the Earth?

As von Baeyer noted, we know there's a three-hour time difference between New York and Los Angeles, and the distance between those cities is about 3,000 miles. Three hours is one-eighth of the 24-hour day that the Earth takes to make one complete rotation, so 3,000 miles is about one-eighth of the way around the globe. Eight times 3,000 is about 24,000 miles around the Earth. The true value is 24,903 miles around the Earth at the equator.

Another such Fermi problem posed by von Baeyer: How much rubber is worn off your auto's tires per revolution? If you assume that a tire with a one-quarter-inch-deep tread is worn smooth in about 40,000 miles, the answer is that a layer of rubber a few hundred thousandths of an inch thick—about the thickness of a molecule—is lost per revolution of the wheel.

John A. Adam, who teaches mathematics at Old Dominion University at Norfolk, describes a number of Fermilike problems in the current issue of *Quantum*, a magazine designed for bright high school math and science students.

One is a problem that Temple University mathematician John Allen Paulos sometimes poses for his students: Estimate how fast, in miles per hour, human hair grows.

Adam's approach is to assume hair grows about an inch between haircuts, and to divide that by the amount of time between haircuts. The approximate answer, when converted, is a few hundred millionths of a mile per hour.

And that, Adam calculated in another problem, is about the same as the rate of growth of a child from birth to 18 years.

Who says that mathematics is remote from the real world?

BRAIN, BODY AND MIND

What is Matter?—Never mind.
What is Mind?—No matter.
PUNCH, vol. xxix, p-19 1855

Some scientists have described the three and one-half pound human brain as one of the most complex assemblages of matter in the universe. During my science writing tenure at the Times-Dispatch, scientists have tried to understand that complexity by analogy with the currently most complex achievements of humans—from telephone switchboards to computers—and to date, such analogies have fallen short of the task. The brain appears to be unlike anything humans have devised, and the more scientists understand about how it really works, the more accurate the scientists' description of its complexity seems to be.

Still unanswered are the fundamental questions about the relationship between the brain and mind, about the sources of creativity, and about how the body—a remarkable, complex assemblage of matter in its own right—can be made ill or well by something as insubstantial as a thought.

Also included in this section are some appreciations of the human body, both the servant and the master of the brain.

What Is It That Governs the 'Part Which Governs'?
OCTOBER 16, 1988

"This being of mine, whatever it really is, consists of a little flesh, a little breath, and the part which governs," wrote Marcus Aurelius Antonius in his *Meditations* about 19 centuries ago.

It's the part that governs that's now one of the most avidly investigated areas of modern science. The part that governs is our brain, which some scientists have described as the most complex three and one-half pounds of matter in the universe.

It's in that connection that a report, in a recent issue of the journal *Science*, about turning on aggressive behavior in lobsters, is of interest.

As a matter of perspective, scientists in various fields are trying, in one way or another, to figure out how the human brain works.

But the human brain is so complex that scientists cannot possibly understand, at least not at the present state of their knowledge, how it works as a whole. They have to work on specific little aspects of it at a time.

And, obviously, they can only investigate the living human brain in extremely limited ways. So they investigate specific phenomena in various animals, in some cases, animals that are farther down the evolutionary scale than humans or primates or even mammals.

The simpler the system, the better it may be for learning some basic lessons about our own "part which governs."

So scientists study such unlikely creatures as squid and lobsters to learn something about our own brains.

Broadly speaking, what scientists have learned so far from such studies is that the brain works through certain principles of chemistry and electricity.

There are billions of nerve cells in the brain, and those billions of cells communicate with one another through tiny electrical impulses that are passed along from cell to cell by particular chemicals.

We know intuitively, if not consciously, that our brains can be influenced by chemicals. There are agents that make us feel good, feel bad and lower our usual guards to proper behavior. "Mind-altering drugs" is a phrase that has become part of our language within the past 30 years.

We also realize that substances within our own bodies can affect our perceptions and our behavior. Adrenaline can key us up. Certain hormones affect our sexual drives.

It's the question of how, exactly, certain chemicals cause changes in behavior that some scientists are trying to answer. And it's a central question in the studies of lobsters that Dr. Edward A. Kravitz has been conducting. Dr. Kravitz, director of the program in neuroscience at the Harvard Medical School, reported on the studies in the Sept. 30 issue of *Science*.

The behavioral results of his investigations are interesting. Certain substances, he finds, can prime the lobsters' bodies to behave in predictable ways, as if they are in certain situations. The situations involve fights for dominance. When two lobsters are put into a tank, they lock claws, push and shove and engage in other kinds of fighting behavior that continues until one becomes dominant, according to *Science.*

The dominant one assumes a "standing tall" posture: It walks on the tips of its walking legs, tucks its abdomen down and extends its claws in front, open. The subordinate one assumes a "standing low" posture, squatting close to the surface, claws to the side and abdomen flexed to an extreme upward position.

Dr. Kravitz found that he can evoke the aggressive, "standing tall" posture in lobsters by injecting them with a substance known as serotonin, and he can elicit the submissive, "standing low" posture by injecting them with a substance called octopamine, which is chemically akin to a substance in our bodies called norepinephrine.

Those substances are important for the communications among nerve cells. The chemicals help pass the impulses from one cell to others.

Serotonin and octopamine, according to Dr. Kravitz, seem to act at various levels in the lobsters' nervous systems, and they incline the creatures' behavior toward certain stereotypical responses.

Dr. Kravitz's work is only one example of many studies in recent years and decades that tend to show that our brains and central nervous systems are as subject to the principles of chemistry and electricity as the motions of the planets are to the principles of physics. Such studies suggest that behavior can be influenced in predictable ways with chemicals and with electrical stimulations of particular parts of the brain.

So, "the part which governs" is itself governed—by natural laws.

But to what degree? That's a question that philosophers have been asking for a long time. Do we not have the willful freedom to govern "the part which governs"? Or are we merely sophisticated automatons with illusions that we can govern "the part which governs"?

More than 100 years ago, Fyodor Dostoevsky considered similar questions in his brooding story, *Notes from the Underground.* He railed against notions of people without choices. If that were the case, he said, we would be nothing more than mechanistic organ stops. He suggested that he himself would, and could, act irrationally deliberately just to show that his behavior cannot be predicted.

One wonders how much choice a thinking lobster would have in not assuming an aggressive posture when its nerve cells are bathed in serotonin.

Wonder of Universe Sits in Our Heads
NOVEMBER 3, 1991

Considering the subject, there was an incongruous air of ordinariness about it.

In a University of Maryland zoology department laboratory, in aluminum baking pans like the ones stowed in many homes' kitchen cabinets, lay two whole, preserved human brains. The organs were whitish, solid, wrinkled, about the size of grapefruit.

They were being shown to some journalists attending a two-week course on the brain and on scientists' efforts to understand it. The course was sponsored by the Knight Center for Specialized Journalism at the university's College of Journalism.

Nothing is known about the persons to whom the brains belonged, explained Dr. Arthur N. Popper, associate director of the university's Center for Neuroscience. It's not known whether they were male or female. It's not known when or how they died. It's not known how they lived.

Presumably, he said, their bodies had been donated for study or research; in many cases, such bodies are unclaimed ones.

The journalists were divided into two smaller groups that gathered around each of the two pans, to receive some real-life lessons on the brain's anatomy.

To me, the experience was something like viewing a Necker cube, which is a line drawing of a cube that changes aspects as you look at it. First, you seem to be looking down at the cube, then as you stare at it, you appear to be looking at its underside.

Similarly, the brain before me alternately changed aspects.

On the one hand, the brain that a graduate student showed us was simply a thing, an object with a well-defined structure that's generally the same in everybody.

On the other hand, it once was more than a thing. It was, through its complex interworkings, the instrument of awareness for somebody who once lived, who walked, talked, felt good, felt bad, laughed, cried, experienced the warmth of sunshine, the cold of night, the comfort of friends and family and perhaps the emptiness of loneliness.

The graduate student points to the frontal lobes, the parts of the brain just behind your forehead, above the eyes.

It was in there that the person to whom the brain belonged once planned what he or she would do tomorrow, or next week, or next year.

The student points out the temporal lobes, on each side of the brain, that in life are located around the ears and toward the front of the head.

It was in there that the person's strong emotional memories resided. That's where his or her memories of first love were registered; it's where, according to one reference, any pangs of jealousy may have surfaced.

The student points out the cerebellum, a pair of structures at the lower back of the brain. In life, the cerebellum helped guide and modulate the person's movements, especially those requiring fine and intricate maneuvers like threading needles or buttoning shirts.

The student points to areas on the surface of the brain that are involved with movements, with the senses, with speech. For the person in whom the brain existed, those were the areas that were responsible for his or her walking, for working with hands and arms, for experiencing pain, for the pleasure of a tickle, and for putting thoughts into words.

Now, that brain is a thing to illustrate a laboratory lesson.

Once, for somebody, it was everything.

In subsequent days, the journalists were to learn what scientists know about how the brain works in life. It ultimately comes down to billions of nerve cells, each connected to many others, yielding numbers of connections so large that scientists express them as 10 multiplied by itself more than a dozen times.

Every sight, sound, taste, smell, touch, thought, idea or feeling involves tiny voltages passed along from nerve cell to nerve cell, with the aid of special chemicals and electrically charged atoms of sodium, potassium and calcium. Some serious ailments, such as Parkinson's disease, result when the cells making just one of those special chemicals die.

But scientist after scientist readily admitted that despite their growing knowledge about how the brain works, their level of ignorance is overwhelmingly greater than their understanding.

During the first several days of the journalists' course, a major conference of astronomers from around the world was being held in the same building. They were discussing some mysteries of modern astronomy, such as quasars, black holes and other strange entities that may lie at the centers of galaxies.

The thought occurred: The most complex structures in the whole universe probably rest within the assembled astronomers' own heads.

Body Is Worth More Than Sum of Its Parts
MAY 11, 1986

Among the overworked cliches that invoke the authority of science is the one that says the materials of our bodies are worth only about 98 cents.

Actually, most of us could not afford to buy ourselves if we had to pay catalogue prices for the various enzymes and other wondrous substances that make our bodies function so well throughout the average human lifespan.

As a Yale biophysicist recently learned, our worth is almost priceless when one's body goes through the shop, so to speak, for some work involving a tiny missing part.

He is Dr. Harold J. Morowitz, who for many years has been writing charming and thoughtful essays in the monthly publication, *Hospital Practice*. A decade ago, he wrote that the worth of his body—or anyone's body, for that matter—was at least $6 million, according to then-current biological supply catalogue prices of various substances that our cells produce.

His latest evaluation comes from a recent accident that amputated the tip of his right index finger. The accident involved a slamming door.

He went to the emergency room; some surgery on the fingertip was subsequently performed and he later had some more surgery done on an outpatient basis. The final bill was $4,731.43.

That led him to thinking about the worth of the human body in a way that was different from the catalogue method.

Dr. Morowitz figured that he lost about 1.3 grams of tissue from the end of his finger. He divided 1.3 into $4,731.43 and found that, on such a scale, his body is worth $3,639.56 a gram. He weighs 85,000 grams, so the value of his body on that scale turns out to be more than $300 million, "a very satisfactory upgrade from the former lowly $6 million" that he had computed a decade earlier.

Actually, someone else has taken a more recent look at the catalogue prices of some of the materials that are in our bodies, and found that only a small fraction of the dozens and dozens of substances inside us are worth around $170,000. Such substances are used in research and in therapy; that's why there are supply firms, with catalogues, that list the prices of so many of the materials in our bodies.

Dr. Daniel A. Sadoff, a Seattle veterinarian, reported in a letter in a 1983 issue of *The New England Journal of Medicine* that the cholesterol in an average 150-pound person's body had a wholesale value of $525. The average person's supply of fibrinogen, which is essential to the blood clotting process, was worth about $740 in 1983 catalogues.

Dr. Sadoff said the hemoglobin in our blood was worth nearly $2,600. Albumin was worth about $4,800; our prothrombin was worth about $31,000, and our immunoglobin was worth about the same amount. A substance that helps oxygen circulate in muscle tissue was worth about $100,000, in the quantities that are in our bodies. His calculations do not include such items as blood. The charge then for blood by blood banks was around $150 per unit, which made the average person's total blood supply worth about $1,200, according to Dr. Sadoff.

One is tempted to raise the analogy of the price of the parts vs. the price of the finished product; a $10,000 car may cost $25,000 to $30,000 if you buy it piece by piece from the parts counter.

But that analogy doesn't really fit. For one thing, one can't really make a working human body by buying parts and putting them together. It's not the parts, it's the integral whole that's important, and no one knows exactly how to provide such exquisite orchestration to a bunch of expensive, sophisticated chemicals.

The 98-cent figure that's often quoted as the value of the human body's ingredients probably refers to the basic chemicals—the carbon, hydrogen, oxygen, nitrogen, zinc, iron and others—that make up the molecules of the working substances in our bodies.

But that's like valuing a painting by calculating the price of the paints and of the canvas, or a computer by the price of the sand used in making the silicon chips used in the circuits, or a sculpture by the price of the clay, granite or marble. The value of such things is not based on the price of the raw materials involved, but in how those materials are organized. We buy a painting not for the paint and canvas, but for how the paints are organized on the canvas.

We pay for how things are put together. And so does nature, whose currency is energy. The more orderly the structure, the higher its price in terms of energy spent. Nature had to spend a considerable amount of energy to produce the elements, such as carbon, hydrogen, nitrogen and oxygen, which are the key ingredients of the molecules in living tissue. Making orderly molecules out of atoms also requires energy.

Living things are constantly paying a price, in the energy from the food they consume, for the orderly processes and structures that keep them alive. (The ultimate bank for such energy, incidentally, is the sun, which showers the Earth with its currency.) In death, the price is no longer being paid, and the orderly structures disintegrate, and the processes stop.

So, in life, our bodies are certainly worth more than the value of the basic elements that are in it. By various measures, our bodies are nearly priceless.

Human Body: What If It Went 'All Over'?
JULY 24, 1983

There is a scene in the movie, *The Wizard of Oz*, where the Scarecrow's straw stuffings are scattered all over the surrounding ground by some catastrophic event.

When the Scarecrow surveys the scene and frantically complains about his plight, while desperately grabbing bunches of straw to stuff inside his body, one of the other characters admonishes him, "Oh, that's you all over" (or words to that effect).

Somehow, that image of being spread over a large area came to mind recently while looking at some figures about the human body. Within the human body are many structures, coiled, folded and intertwined, that could cover some unbelievable distances and areas if they were uncoiled, unfolded or unraveled.

Unraveling, in fact, suggests a more apt image, namely, that of a ball constructed of hundreds of feet of string. When the ball, perhaps just several inches in diameter, is unwrapped, its contents extend across a football field or two.

Unravel a human body, and it could extend around the world and, indeed, into the solar system. The figures associated with the human body are astronomical in more than one sense of the word.

One human body, for example, contains trillions of cells. If each cell in a typical human body were matched one-for-one with the stars in the Milky Way galaxy, one would have to go to another galaxy and borrow at least a few hundred billion more stars to complete the job.

Or, take the amount of DNA in each cell's chromosomes. DNA is a coiled molecule that contains the genetic instructions that are passed along to subsequent generations of cells, and of human beings.

If the DNA from the chromosomes of one cell were uncoiled and laid end to end, it would stretch out to about one meter, or slightly more than a yard. There are trillions of cells, so if all of the human body's DNA were stretched out and laid end to end, the DNA line would stretch a billion kilometers—or hundreds of millions of miles.

It would extend at least from the sun to Jupiter.

Less dramatic is the total end-to-end length of the blood vessels in the human body. That figure is generally said to be between 60,000 and 70,000 miles.

Thus, the blood vessels in one human body would extend, if placed end to end, around the Earth more than two times.

Then, there are the internal structures of the kidneys. Each kidney weighs about a half a pound, and is no more than five inches long, three inches wide and two inches thick. Within each kidney are structures called tubules, through which most of the blood's fluid is reabsorbed after being filtered.

The length of those tubules in both kidneys add up to nearly 40 miles.

If we were to unravel various components associated with the skin, we again come up with some dramatic figures about the body's potential extension in space.

According to an American Medical Association booklet about the human body, the average body contains about 3,000 square inches of skin. Each square inch, according to that booklet, *The Wonderful Human Machine*, contains about 15 feet of blood vessels. That works out to between eight and nine miles of blood vessels in the skin alone.

Those miles of blood vessels in our skin, incidentally, help perform an important function during heat waves and cold snaps. The skin is a temperature regulator, and temperature regulation is carried out to a significant extent by the blood vessels in our skin.

When the body must lose heat, those vessels open up to allow more blood to flow through them, permitting heat to be dissipated. Moisture secreted by myriad sweat glands in the skin also are part of the heat-reducing mechanism. When the body must conserve heat, the vessels narrow to reduce the amounts of heat lost through the skin.

In addition to its rich endowment of blood vessels, each square inch of skin contains 72 feet of nerves, or more than 40 miles' worth in our entire covering package. The skin of the average adult, incidentally, weighs about six pounds.

Deep inside our bodies exists yet another structure with the potential for extensive coverage in space. It is within the small intestine, which is about 10 feet long.

But lining those 10 feet of intestine are myriad tiny projections, giving the intestinal lining an appearance that one author describes as being like a fine Turkish towel. Those tiny projections create a total surface area inside the small intestine, where the bulk of food absorption takes place, of more than 5,000 square feet, or about a tenth of an acre.

So that's the human body all over. It's a remarkable design that can compress a yard of DNA in each cell, 60,000 or 70,000 miles of blood vessels throughout, miles of tubules in each kidney, miles of nerves and blood vessels in the skin, a fraction of an acre within a 10-foot section of intestine, and enclose the whole thing with six pounds of skin.

Sometimes when we are inclined to feel small, perhaps we should think of just how big we potentially are.

Illness Can Stem from Contagious Empathy
APRIL 24, 1983

It's as contagious as a thought.

And while it may affect a number of people at about the same time, with symptoms that are at the least disturbing, the ailment generally passes rather quickly.

But that's not necessarily true of the investigations that follow. Something made a number of people ill, in the same place and at about the same time. Yet, the search for a cause—perhaps a toxic chemical, gases from a faulty furnace or something spread by a saboteur—generally is unproductive.

And when the probable explanation of the outbreak is pronounced by the medical investigators, it may be greeted with considerable suspicion, if not disbelief. The explanation may be that the outbreak was due more to psychology than to physical agents, to the contagious empathy of seeing other people become ill. The disbelief comes from accepting the notion that an illness with real, physical symptoms can spread through a group on something as insubstantial as empathy.

"Mass hysteria" is the term most often applied to such outbreaks, and they may not be particularly uncommon.

In recent weeks, a possible series of such outbreaks contributed to the tensions in the Middle East. More than 800 persons in the West Bank, nearly all of them Palestinian high school girls, have gone to hospitals complaining of stomach aches, nausea, blurred vision and dizziness.

There were allegations that the ailments were due to an Israeli plot to poison Arabs, but Israeli officials denied it. Israeli physicians claim that extensive tests revealed no physical basis for the outbreaks, and suggested that the Arab-Jew tensions in the area may have contributed to the ailments. Meanwhile, some independent investigations by international organizations are underway.

There have been similar cases closer to home. On December 14, 1981, some students at a Newport News high school began becoming ill in the morning, followed by more illnesses after lunch. A cascade of illnesses occurred in the cafeteria. Some who fainted were assisting others who had fainted. The building was evacuated and ventilated by the fire department; carbon monoxide from a faulty furnace was initially suspected.

At a nearby armory, where unaffected students were taken, some more became ill. The most commonly reported symptoms, according to Dr. Tom A. Sayvetz, a state Health Department epidemiologist who investigated the outbreak, included headaches, dizziness or lightheadedness, hot and cold chills, muscle weakness, nausea, fainting and rapid or difficult breathing. A total of 56 persons were affected and nearly all were students, mostly female students.

Besides the state and local health departments, the Air Pollution Control Board and the state's Occupational Safety and Health Program were involved in the investigations of the incident. The preliminary environmental and medical data failed to establish a physical cause of the problem.

More recently, an outbreak among some elementary school students in Massachusetts was reported in *The New England Journal of Medicine.* Dizziness, weakness, headache, nausea, abdominal pain, hotness and coldness, and fainting were among the symptoms. Environmental studies revealed nothing unusual, but laboratory studies indicated low quantities (in the parts per billion range) of a chemical in the urine of those affected.

The chemical is one that's normally associated with plastics, some insecticides and disinfectants. The investigators concluded, after some studies, that it came from the plastic containers in which the specimens were contained, or from the plastic tubing of the analysis instrument, and was not related to the outbreak.

There are, as the authors of the *New England Journal* report noted, several features of outbreaks due to mass hysteria that distinguish them from those with physical causes. For one, laboratory and other studies fail to turn up a physical cause of the outbreak. For another, the preponderance of illnesses tend to occur in girls or women. If a physical agent were involved, one would expect males to be affected as often as females.

Also, the illnesses in a mass hysteria outbreak tend to be transmitted by apparent sight or sounds. Occurrence of outbreaks among adolescents or preadolescents may be another clue, as would a rapid spread and rapid remission of the illness. "Whenever people under stress gather together, the potential for mass hysteria exists," said Dr. Gary W. Small and Dr. Jonathan F. Borus in their *New England Journal* report.

And when such outbreaks occur, they may raise some controversy about whether they were caused by physical or psychological factors. In fact, psychological factors can cause bodily disturbances that are just as real as those caused by germs of chemicals.

Creativity Flourished As Mental Powers Declined
MAY 2, 1996

Curiously, while his mind was going, his creativity flourished.

As dementia gradually robbed him of his mental functions, the American abstract expressionist artist Willem de Kooning suddenly started painting again, turning out more than 250 paintings in five years.

What can such an outburst tell us about creativity and the failing mind? wonders Dr. Carlos H. Espinel. He's a Northern Virginia physician whose avocation is studying art for medical insights, either in artists' subjects or in the artists themselves. His analyses are occasional features in the British medical journal *Lancet.*

Espinel, who received specialty and subspecialty training in the late 1960s at Virginia Commonwealth University's Medical College of Virginia, is director of the Blood Pressure Center of Washington, located in Arlington, and he has teaching appointments at the Georgetown University School of Medicine.

De Kooning, according to Espinel, began showing the earliest signs of forgetfulness and deteriorating mental capacity in his early 70s, about

20 years ago. He was forgetting people's names and recent events. (De Kooning died in March 1997 at age 92.)

De Kooning, Espinel said in last week's issue of *Lancet*, tried to cover his growing confusion with small lies or wisecracks. He apparently knew his intellectual abilities were slipping and sometimes he raged against the decline.

Gradually, his output of paintings diminished and finally stopped by the time de Kooning was in his mid-70s. "Alcohol, prescription drugs and depression were the companions of his solitude," Espinel wrote.

He said a number of factors probably were behind de Kooning's decline in mental abilities, including artery disease, alcoholism, uncertain nutrition over the years, medicines, depression and Alzheimer's disease.

Espinel said the world mourned the loss of one of the great 20th-century painters as de Kooning's dementia progressed. ``His prognosis was dismal,'' Espinel said.

But within the next few years, de Kooning started painting again, with considerable vigor and productiveness. "If in 1980 he completed only three paintings, from 1981 to 1986 he finished 254," Espinel said.

One of those paintings, "Untitled III," in the National Gallery of Art in Washington, was completed in 1986 when de Kooning was 82, when he was showing advanced signs of what doctors had diagnosed as Alzheimer's disease.

Such an outburst of creativity during the period when his mental decline was in progress "has fascinated me," said Espinel by phone. "It was a new style, in a way."

During that period, his physical health improved through the efforts of his wife and a group of friends. With their support and encouragement, he stopped drinking, he began eating balanced meals and he exercised every day. And he returned to painting.

"His rehabilitation technique was unique," Espinel said. "He borrowed colors and forms from his own work. He scattered his early masterpieces about his studio, consulted their photographs and projected their images onto the empty canvases. Like a beginner, he copied and retraced images.

"Then he let himself go."

Espinel describes de Kooning's newer abstract paintings as having joyful and balanced colors and forceful lines, sometimes with a sense of movement to them. It all indicates, Espinel said, that "he was making choices and exercising judgment."

Could it be, he wondered, that de Kooning sought to restore his fading sense of self through form and colors, the once familiar elements of his expression?

SEASONAL

Calendars to the contrary, nature's year begins with the spring.
EDWIN WAY TEALE, *North With the Spring*

The Earth, spinning like a top, revolves around the sun. The axis of our spinning-top-Earth runs through the North and South Poles, and that axis is tilted with respect to its motion around the sun. So part of the year, our North Pole leans toward the sun and part of the year, it leans away. The patterns of our temperatures, the lengths of our days and nights and our climates are influenced by that rhythmic nodding of our hemisphere toward and away from the sun, and life itself becomes deeply attuned to cycles that are triggered by a seemingly straightforward astronomical fact.

From frogs to charged fingertips, phenomena related to the seasons have been of continual interest to me.

Frogs May Yield Clues for Preserving Organs
MARCH 21, 1993

Like the residents of Dr. Seuss' *Who*-ville, who shouted in unison so Horton could find them among the clover leaves, spring peepers in Central Virginia are beginning to sing out the fact of their existence.

One is tempted to believe that they are proclaiming, like the tiny people of *Who*-ville (a place the size of a speck of dust), "We are here! We are here!"

The fact that they are here, and are singing so loudly about it as spring begins, is still somewhat of a scientific mystery. Like several other kinds of frogs, spring peepers can freeze during the winter months and thaw, apparently no worse for wear.

Scientists still don't know precisely how they do it.

If they did, they might have a better idea than they have now about how to preserve organs for days or weeks, or perhaps longer, for transplant operations, thus opening up the possibility of organ banks.

To carry the speculations further, if scientists knew all of the secrets of the spring peepers and some of their cousins, they might have some insights into how to put people into states of hibernation.

The reasons might be medical—some treatments or surgery might go better if conducted on patients whose metabolism has slowed almost to nothing. Or the reasons may be more esoteric—hibernation may indeed be a key to sending people out on long space journeys, just as many science fiction plots suggest.

Scientists have been working for some years on understanding how spring peepers, wood frogs and certain other creatures can survive freezing. A little more than a decade ago, William D. Schmid at the University of Minnesota collected several species of frogs, spring peepers included, and slowly cooled them under controlled laboratory conditions to around 21 degrees Fahrenheit. Water freezes at 32 degrees.

He kept them frozen at that temperature for five to seven days; they showed no evidence of limb movement during that time, Schmid reported in the journal *Science* in 1982. In the case of spring peepers, more than one-third—36 percent—of their body water froze.

Kenneth B. Storey and Janet M. Storey at Carlton University in Ottawa, Canada, in a recent summary of what scientists know about how some animals survive winter freezing, note that a number of hibernating reptiles and amphibians can endure days or weeks with as much as 65 percent of their body water locked up in ice.

The main trick, Schmid said in a telephone interview, is to keep the ice outside of cells. And various substances, including glucose, help draw water out of cells, dehydrating them and preventing sharp, damaging ice crystals from forming within those delicate structures.

One way of performing that trick, according to the Storeys, is for the creatures to produce ice-nucleating proteins, which are substances in the blood that trigger and control ice formation in the fluids that bathe the cells.

Another means of keeping ice from forming inside cells is for the animal's system to produce glucose, a sugar, that helps dehydrate the cells. Spring peepers especially make use of that strategy.

As ice crystals start forming on the frog's skin, glucose is released from storage in the liver, with the highest levels accumulating in such vital organs as the liver, heart, brain and kidney. When the frog thaws, the glucose is returned to the liver for storage.

"Frozen frogs have icy skins, stiff limbs and opaque eyes," wrote the Storeys in the *Annual Review of Physiology*, 1992. "Internally, ice crystals run under the skin, between the skeletal muscles and a mass of ice fills the abdominal cavity. Internal organs appear shrunken."

When the temperature warms, the crystals disappear and shrunken organs enlarge to their normal sizes.

And during early spring evenings, the spring peepers sing their songs of survival, like loud and restless wind chimes.

Practical Ben Franklin and Daylight-Saving Time
APRIL 5, 1992

Today is a day that Benjamin Franklin would have appreciated.

It is the beginning of this year's daylight-saving time season, the day we set our clocks forward one hour to "give" us an extra hour of daylight for outside activities. Other advantages, proponents say, include cutting down on our light bills and on traffic accidents associated with night driving conditions.

It is, of course, a mere accommodation. We haven't affected the measurement of time in any way, except to slide it forward by an arbitrary amount.

And it doesn't suit everybody. As one farmer has been quoted as saying in the past, the cows still give milk according to "God's time," or to the cycle to which farm life is attuned. The hay may be dry enough to be worked by 10 a.m. on standard time, but not by 10 a.m. daylight time.

Such, however, were not the concerns of Ben Franklin, who more than 200 years ago suggested the value of using daylight to its fullest. Being a frugal man, he based his argument solely on issues of economy. "I love economy exceedingly," he wrote in a 1784 essay titled "Daylight Saving," that appeared in *The Journal of Paris*.

It was a tongue-in-cheek essay that poked fun at Parisians and others who stayed up late at night and wasted the first hours of daylight sleeping. He presents himself, in fact, as such a person (actually, Franklin was an early riser) who is astounded to discover that light flows through his shuttered windows around 6 in the morning.

A natural philosopher, he claimed in the essay, tried to convince him that it really wasn't light coming in the window, but darkness escaping because he had failed to close the window's shutters.

Franklin complained that too many Parisians were trading daylight hours for the candlelit life, and daylight, after all, is far cheaper than the light provided by candles.

Figuring that there were then about 100,000 families in Paris, and that on average, each family burned half-a-pound of candles per night, he concluded that the city's dwellers burned more than 64 million pounds of candles between March 20 and Sept. 20. When multiplied by the currency of the day, Franklin concluded that an immense sum was being consumed in candle wax and tallow "that the city of Paris might save every year by the economy of using sunshine instead of candles."

Further, he said the wax and tallow that would not be consumed during the spring and summer would mean that candles would be cheaper during the ensuing winter.

But Franklin's essay on the advantages of saving daylight apparently went unheeded until the early part of the 20th century, when a London builder named William Willett published a pamphlet in which he complained about the daylight hours wasted under the standard time system. He suggested that the system be changed in the summertime so that people could take advantage of the daylight for outdoor recreation, especially family-oriented activities.

And, like Franklin, Willett noted that his daylight-saving scheme would reduce the expense of artificial lighting.

His plan was ridiculed at first, especially by farmers, according to Dr. William Andrewes, a former curator of The Time Museum in Rockford, Ill., and now at Harvard University.

But during World War I, Germany and then England adopted daylight-saving time as an economy measure in 1916, a year after Willett died. The United States adopted the new time scheme in 1918, but Congress repealed it the following year when the war was over.

Then, during World War II, year-round daylight-saving time was introduced again in this country as an economy measure. The so-called "war time" began in February 1942 and lasted until September 1945.

More than 20 years later, Congress passed the Uniform Time Act under which the nation was to observe daylight-saving time, starting at 2 a.m. on the last Sunday of April and ending at 2 a.m. on the last Sunday in October. The act, which went into effect in 1967, allowed states to exempt themselves from the act, and some did.

When the country's oil supply was threatened by an embargo in the early 1970s, Congress put most of the nation on year-round daylight-saving time, again in the interest of conservation, from early 1974 to late 1975.

And in 1987, new federal legislation went into effect that moved the start of daylight-saving time from the last Sunday in April to the first Sunday in April, with it still ending on the last Sunday in October.

Among the reasons for that change, which we observe today, was a Transportation Department estimate that the earlier beginning date would help save millions of dollars in traffic accident costs, not to mention the prevention of injuries and deaths.

Ben Franklin, incidentally, even considered the transportation angle in his 1784 essay.

"Let guards . . . be posted," he wrote, "to stop all the coaches, etc. that would pass the streets after sunset, except those of physicians, surgeons and midwives."

Physics of Mirages, Flying Bugs and Windshields
JULY 13, 1995

It's a puddle you see but never reach.

You are driving on a highway on a hot day and you see a shimmering, shining patch that appears to be water on the road ahead. When you arrive at that spot, nothing's there. But there's another one ahead.

And so on.

The puddles are mirages. They are reflections of the sky on the road and they are one manifestation of scientific principles that apply when we are traveling on the highway, as many of us are during the summer months.

Physics and aerodynamics are hardly uppermost on our minds when we are on the road, but we nevertheless are subject to their uncompromising dictates.

We are moving objects, for example, and therefore subject to the laws of motion, which help explain such things as why our cars need a low gear to get going to why our bodies slide outward when we go around curves.

There are real-world lessons about friction. The good news is that friction enables our cars' tires to grip the road and move us over the highway, as well as to stop. The bad news is that friction is costly; engineers say that about 1 gallon of gas in every 5 is used to overcome friction in the engine and drive train.

One of the more visible and common science lessons in highway travel, according to physicist Richard H. Zallen at Virginia Tech, lies with the mirage of the shimmering puddle ahead.

It's due, he explained, to the hot layer of air just above the road that bends the path of light rays from the sky. Light speeds up a bit when it passes from a dense medium, such as cool air, to a less dense one, such as hot air. It slows a bit going the other way. And when light changes speed, its path also changes.

Zallen used the analogy of a lifeguard in water who has to reach a swimmer in trouble some distance away. His path is bent as he attempts to get to the swimmer as quickly as possible. He swims to shore, runs along the beach and swims out to the person in distress. His speed is reduced in the water, increased on the beach and slowed again when he re-enters the water.

Because light's path changes as it passes from one medium to another, we see things in places where they really aren't, such as the fish beneath the water's surface, the stick that appears bent when inserted into water—or the sky seemingly on a patch of road in front of you.

A person in a car traveling on the road sees the rays re-emerging from the hot layer. To him or her, those rays appear to be coming straight from the road. But if you trace out their paths from sky through cooler air to less dense warm air and to cooler air again, you find their source is the sky.

Meanwhile, the bugs that smash on your windshield during the summer would do well to follow curving paths. Why aren't they carried in the air stream that flows up and over the windshield instead of being smashed into it? Because bugs are more massive than air molecules, explained James F. Marchman, an aerodynamics expert at Virginia Tech. Since bugs are relatively massive, they are less likely to change their direction than air molecules; they are, in accordance with Newton's first law of motion, likely to continue their motion in a straight line.

And continuing their straight-line motion, from the front of the car to the windshield, generally means a kamikaze mission for the bug, a mission that ends in a windshield splat.

Such is the price of ignorance of highway physics and aerodynamics.

Was Chicken or Egg First? Or Was It the Rabbit?
MARCH 31, 1994

This weekend, the humble chicken egg will be a dominant symbol in Easter-related activities.

It will be hard-boiled, dyed, hidden and sought.

It will be imitated in chocolate and other sugary forms.

And, for this weekend only, it will be linked not to a chicken but to a rabbit.

All of which draws attention, scientifically speaking, to one of nature's most wondrous packages.

First of all, there's the geometry of the package, a pleasingly shaped object that gave rise to the word "oval," from the Latin word ovum, for egg. The ideal oval is an ellipse, but the outline of an egg doesn't really meet that ideal.

The ellipse is uniform. The egg's outline has a blunt end and a pointed end. The shape can vary from hen to hen and eggs from a particular hen always have the same shape, according to Dr. Paul B. Siegel, University Distinguished Professor of Poultry Science at Virginia Tech.

"It's almost like she fingerprints her eggs," he said.

And, Siegel noted in a brief telephone lesson on the egg, it's generally a good thing that eggs have the shape they do. It's useful in a survival-promoting way.

"Immediately after the egg is laid, it undergoes rapid cooling—the chicken's body temperature is 104 degrees—and it cools more rapidly on the large end," he said.

The cooling is more rapid at the large end than the small end because there's more surface area to radiate away heat; additionally, there are more pores in the large end of the shell than at the small end.

Because of the rapid cooling, an air pocket forms inside the shell beneath the blunt end of the egg and the inner membrane that surrounds the so-called white.

That air pocket, Siegel said, is important because it allows the chick inside to make a transition shortly before hatching from a liquid environment to breathing in an air environment.

When a chick breaks into that air sac, Siegel said it starts a swallowing action that makes a clicking sound—around three clicks a second. That clicking sound signals other chicks to break into the air pocket and prepare for hatching.

Siegel said the clicking synchronizes hatching; it causes all the chicks in a clutch of incubated eggs to come out of their shells about the same time, about 10 or so hours after the first chick starts making the clicking sounds.

"They have to hatch very close together," said Siegel.

Newly hatched chicks are up and running within a short time after breaking out of the shell; the mother hen could not keep all of her brood out of danger if she were dividing her attention between those that have already hatched and those that hatch later.

So the ultimate usefulness of the shape of the egg is that it helps chicks prepare for the air-breathing world, and in the process, it helps all of the chicks in a clutch of eggs enter the world about the same time. They have a better chance of surviving that way.

"It's kind of neat," Siegel said of synchronized hatching. "Nature's wonderful."

Any discussion of the egg must also include the time-worn question of which came first, the chicken or the egg?

From an evolutionary point of view, noted Dr. Harold McGee in his book *On Food and Cooking*, the egg came millions of years before the chicken. He said that modern fowl and bird eggs are refined versions of the leather-skinned eggs that amphibians first laid on land 250 million years ago.

But he is completely silent on the history of dyed eggs purportedly laid by rabbits.

Inventive Insects Survive Winter in Many Ways
NOVEMBER 16, 1986

Except on an occasional, unseasonably warm day, the insects are gone now.

For the most part, we no longer hear them, see them, shoo them, swat them, dodge them or otherwise have to acknowledge their presence in positive or negative ways.

We probably don't consciously notice or think much about their absence, but subconsciously we know they will be back when spring comes. Insects, after all, are ubiquitous; there are nearly 90,000 different kinds that scientists know of, and more than half of all living things on

Earth are insects.

The question is, where do they go when winter comes? How do they survive?

The places they go include limbs of trees and shrubs and the soil, beneath the autumn and winter cover of fallen leaves and dead weeds and grasses. In such places are the eggs, or the worm-like larvae, or young, miniature versions of some of next spring's populations of insects.

Their winter survival techniques are many. Some, like ants, go a few feet underground to seek moderate conditions. Others, like bumblebees, paper-making wasps and hornets, produce fertilized queens that hibernate over winter while the workers are wiped out by hard frosts and freezes.

As scientists have been learning in recent years, many insects use tricks of chemistry to help them survive.

A number, especially those in the larval stage, apparently load their bodies with alcohol, get rid of water and replace it with fat, empty their digestive systems and settle in for several months of dormancy during the cold weather. With their alcohol antifreeze and other protective measures, they can survive long periods of sub-freezing temperatures.

The adults of some kinds of insects, like flies, get rid of excess body water and seek out a protected place, such as tree holes or in buildings, to survive the winter. The depth of their dormancy is not as great as that of larvae, which means they can be easily stirred when conditions are favorable.

That's why the first warm day of late winter may bring out the flies; they can quickly become aroused from their dormancy.

The state of dormancy that many insects enter to endure winter is called diapause; it is a time of changed, slowed bodily activity.

And preparation for the winter begins in late summer or early fall. It's not triggered by lowering temperatures; indeed, the changes begin when the weather is still warm. Rather, as scientists have learned, the signals that start insects preparing for winter survival are changes in the length of the day.

The insects' internal body clocks take note of the shortening days of late summer or early fall, according to Dr. James Liebherr, an entomologist at Cornell University's New York State College of Agriculture and Life Sciences, and touch off chemical changes.

Particularly affected are the larvae of many insects, whose bodies start getting rid of water. Water freezes at temperatures of 32 degrees

Fahrenheit and lower, and water that freezes within their bodies would be severely damaging if not fatal.

Ice crystals begin forming around impurities, so Dr. Liebherr noted, insect larvae "clear their guts" before settling down for winter. That reduces impurities and thus opportunities for ice crystals to form.

At the same time, the larvae's internal chemistry changes. Their bodies start making an alcohol called glycerol, which freezes at a much lower temperature than water. That's the principle behind the antifreeze you put into your car's radiator; the antifreeze contains an alcohol that keeps your radiator from freezing and becoming damaged.

The lowest recorded temperature at which an insect larva has been known to survive, Dr. Liebherr said during an interview, is 70 degrees below zero Fahrenheit.

Besides putting its own form of antifreeze into its system, Dr. Liebherr said an insect larva also replaces much of its lost water with fat, which provides much of the food the creature needs to survive through its winter hibernation.

Certainly by now, all of those changes have taken place in those insects that use that strategy to survive the coming winter. Based on detailed studies of some species, he said entomologists assume that most insects that live through winters in larval stages do so through such meta-bolic techniques.

Most insects in adult stages do not make it through the winter. The first frosts and hard freezes cause widespread exterminations, with fleas, yellow jackets and hornets among those being wiped out.

Before that occurs to the yellow jackets, wasps and hornets, whose hibernating queens will ensure the species' survival the following spring, the food supplies for them become scarcer and scarcer and the creatures become hungrier and hungrier.

As a result, the yellow jackets and hornets become more aggressive during the autumn, Dr. Liebherr said. Their stings are no more potent than earlier in the season, but the creatures are meaner.

"It's a tough time for them," he noted with a certain tone of sympathy.

Zap! Cold Days Provide Physics Lesson
March 7, 1993

One might say, if one is into bad puns, that the zap has been rising all winter.

A zap is the little crackling spark, accompanied by a sharp jolt, that jumps between your fingers and a metal object, or between your hand and someone else you touch, or between spouses' lips during quick goodbye kisses in the mornings.

The zap is every person's real life introduction to the branch of physics known as electrostatics. It's all about static electricity and it's one of those irksome, indirect consequences of winter that, one hopes, will soon end.

It's not the coldness per se, but the relative dryness of the air—especially indoors where heating systems are often working hard—that promotes zapping. And therein lies a physics lesson.

The zap itself arises from the powerful attraction between positive and negative electrical charges. A negatively charged particle and a positively charged one avidly combine, emitting a bit of energy in the process.

When lots of charges combine suddenly, the results can range from the spectacular flashes of lightning in a thunderstorm to the tiny sparks that people emit when they touch doorknobs, metal cabinets, stair railings, car door handles or other people.

The fundamental unit of negative electric charge is carried by an electron, which is one of the basic particles of atoms. The fundamental unit of positive charge is the proton, another basic atomic particle.

All physical things, of course, are made of atoms. And in all physical things, including us, electrical charges are normally in balance. That is, we are neither positive nor negative, electrically speaking.

But when we put two dissimilar things together, such as the bottoms of our shoes and a carpet, a few electrons will flow from one to the other, at places where the two things are in contact.

Rubbing our shoes over the carpet—or walking across it—causes more of the surfaces to be in contact and thus more electrons flow.

The same sort of thing happens when we slide across a car seat or run a comb through our hair. The jostling contact of clothing against our bodies during normal movements can have the same effect.

The net result is a buildup of separated electrical charges; one thing loses electrons and becomes positively charged, and the other gains electrons and becomes negatively charged.

Under the right circumstances, the built-up charges can discharge through our bodies when we touch a metal object or another person.

The zap occurs.

Physicists say that impressively high voltages—a volt is a measure of the eagerness for positive and negative electrical charges to join—can be involved in those sparks that jump from your finger to doorknobs or other objects. There may be 15,000 or more volts behind it.

(It's not the voltage that's hazardous in electricity, but the actual amount of current that flows that's the problem. Current is measured in amperes. In the case of static electricity zaps, the voltages may be high but the currents are minuscule.)

Weather figures into the static electricity zapping picture through moisture in the air. Moist air tends to keep the charges from building up; the electrical charges continually drain through contact with the surrounding microscopic water droplets.

Cold air can't hold as much moisture as warm air, so cold winter days are ideal setups for becoming zapped when you slide into or out of your car. And the relatively dry air in homes and buildings, resulting from heating systems running so much, can help set you up for zaps when you walk across carpeted rooms.

The same kinds of charge-building processes can occur on large scales in nature. Jearl Walker notes in his book *The Flying Circus of Physics*, that blowing sand and snow can result in things that shock people when they touch them.

He cited reports from the Colorado Rockies about fences becoming charged during blowing snow. The charges can become so great that men and cattle are knocked down when they touch the fences.

Again, it's all due to the transfer of electrons. In this case, the transfer takes place between the snow particles and the metallic fences.

Obviously, nature has many clever ways of setting up the zap.

WHEN WE WENT TO THE MOON

That's one small step for a man, one giant leap for mankind.
NEIL A. ARMSTRONG, *first words spoken as he stepped onto the moon July 20, 1969*

Okay, let's get this mother out of here!
EUGENE A. CERNAN, *last words spoken on the moon at liftoff of Apollo 17 lunar module, December 14, 1972*

The following four pieces are what are known in the newspaper business variously as "analyses," "commentaries," "think pieces" or "thumb suckers"—pieces that take a look at developments or trends, place them in some sort of perspective and suggest what they may mean. Each of the following was written under deadline, immediately after the conclusion of a significant moon-bound mission. They included Apollo 8, which was the first time that human beings traveled to the moon (but didn't land), and were far enough away from the Earth to see and photograph our planet as a beautiful blue ball in the blackness of space; Apollo 11, the moon landing mission; Apollo 13, the mission that summoned human history's largest collective rescue effort, and Apollo 17, the final moon-landing voyage..

Apollo 8 was my favorite among the moon missions. It was the Jules Verne voyage to the moon come true. It occurred at Christmas time in 1968, and it was a positive, proud ending to an otherwise horrible year for the United States that brought the assassinations of Robert Kennedy and Martin Luther King, Jr.; the chaotic and confrontational Democratic National Convention in Chicago, and continuing internal upheavals over the war in Vietnam. Apollo 11 was memorable, of course, because it was the first time that human beings landed and walked on a celestial body other than Earth. The piece on Apollo 13 focuses on the night the near tragedy occurred—the mood, confusion and fears that permeated the Manned Spacecraft Center in Houston during the first few fearful hours of the accident. Then, with the suc-

cessful splash down of Apollo 17 more than two and a half years later,
the moon landing program—born in the hottest part of the Cold War with
the Soviet Union—came to an end, and so did one of humanity's boldest
exploration adventures.

Apollo 8 Viewed As Milestone in History of Mankind
DECEMBER 29, 1968

HOUSTON—Throughout the past week at the Manned Spacecraft Center here, there was a touch of Christmas in the air and a profound sense that history was being made by the flight of Apollo 8 to the moon and back.

It was not the sense of history that applies to most of the other news events of last week, the type of history that flares up in newspaper headlines for a few days, then dies with time.

Rather, the journey of Apollo 8 was being viewed as something that will make a difference in the story of humankind, equivalent to such past milestones as the first controlled nuclear reaction, the first time man sent a radio signal, the first time he flew, and the first time someone sailed far enough to prove the Earth did not have an edge to it.

The voyage of Apollo 8 actually represented a series of triumphs, according to space officials and space exploration observers here.

First of all, it was technological triumph. For example, the Saturn V rocket that launched Apollo 8 and its crew of Frank Borman, James Lovell and William Anders, has more than five million functional parts. If it were 99.9 percent reliable, there would be more than 5,000 defective parts.

Yet, so far as could be determined here by week's end, the launch phase of the mission was textbook perfect, demonstrating a reliability greater than 99.9999 percent.

The Apollo 8 spacecraft itself has more than two million functional parts, not counting such things as wiring. As far as mission control here could tell, there was not a single, major malfunction throughout the six-day, half-million-mile voyage into the nonforgiving regions of space.

There was a mental triumph. Mathematicians, physicists, astronomers and computers had figured the flight out on paper a long time before Apollo 8 actually flew. Velocities and distances all the way to and from the moon had been computed, the effects of the moon's and the

Earth's gravity fields were figured.

The immensely complex problem of launching a spacecraft from a given point on Earth, which is spinning, wobbling and moving through space, to hit a given point near the moon, which also is spinning, wobbling and moving relative to the Earth, was theoretically solved.

Just how well such problems were solved was demonstrated by Apollo 8's actual flight trajectory, which required only one or two minimal midcourse corrections going to and returning from the moon.

And everything happened according to schedule. Over a mission that lasted a total of 147 hours, the total actual variation of major events was only a few seconds off.

The total feat may be likened to, or reduced to the efforts of submicroscopic creatures, living beneath the skin of an apple, to throw a tiny dart through the skin and hit a moving ping-pong ball seven or eight feet away.

But the apple must be spinning and moving, and so must the ping-pong ball. The apple analogy may even be carried a bit further by way of noting still another triumph of the voyage of Apollo 8. The skin of the apple represents the Earth's atmosphere, and most of human history has been carried out just beneath such a thin skin.

Even when man first began Earth orbital missions, he was only an infinitesimally small distance from the outside of the skin. But with last week's Apollo 8 mission, three men have told and shown millions of their fellow human beings what they saw and experienced while doing so.

One space science consultant, who observed the Apollo 8 mission from here during the week, believes the flight cannot help but have a profound impact on man's perspective of himself.

He further noted that 3:30 p.m. EST Monday marked the first time that human beings—specifically Borman, Lovell and Anders—would have fallen freely on another celestial body, rather than on the planet Earth.

This was the point when the Apollo 8 spacecraft entered the moon's sphere of gravitational influence; this was the period when the Earth's gravitational field no longer dominated the spacecraft.

The mission also marked the first time that human beings have been able to see a whole side of the Earth—at least, the lighted portion— at one time.

It was also the first time that the Earth stood out for human beings as a beautiful, comforting planet in contrast to an extremely desolate, barren celestial body. As Lovell put it, "The Earth from here is a grand ovation to the big vastness of space."

That bit of spontaneous poetry has become one of the most quoted lines here from the lunar mission.

And there were still other levels of significance to the mission. "The Russians can land men on the moon tomorrow, and it won't make any difference," said one person here last week. "We've scored a first, maybe our biggest first in space."

In addition, there were possible scientific triumphs in the observations of the astronauts as they circled the moon 10 times from about 70 miles above its surface. They reported dark spots within craters on the moon's back side; they reported on a puzzling fine white haze that appears over the moon's limb about two minutes before the sun rises there, and they found some craters that "definitely suggest volcanic activity."

Besides their observations, the more than 1,000 photographs—black and white and color—and the motion pictures they took of the moon will undoubtedly provide grist for scientists' mills for months and possibly years to come. These photographs are expected to be many times sharper and clearer than any that unmanned spacecraft have relayed back from the moon.

The general feeling here when Apollo 8 successfully splashed down in the Western Pacific Friday was that the voyage of Borman, Lovell and Anders was not simply a fantastic voyage, but one that—as with past great voyages in history—expanded our horizons and changed our spirit.

Meanings of Apollo 11 Complex
JULY 27, 1969

HOUSTON—Science fiction became science reality last week when men in fact journeyed to the moon and loped across its barren, alien landscape with unearthly grace and agility.

Most of the world watched the event on home television screens—itself a magical, awe-inspiring feat. True, the TV pictures were not of the slick, sharp quality of Earth studio transmissions. Nevertheless, the impressions of those two and one-half hours of live scenes from the moon last Sunday will probably be as lasting as the footprints Neil A. Armstrong and Edwin E. Aldrin, Jr. left on the moon.

Many people have seen similar scenes in good and bad science fiction movies—scenes of spacemen dressed as spacemen should be dressed, with their spaceships in the background.

But this time, the scenes were real, and that fact touched off ripples of exhilaration, pride, awe and fascination in a global mind.

The successful landing of a spaceship named Eagle on the moon last Sunday afternoon and the subsequent walks of Armstrong and Aldrin on the lunar surface have a multitude of meanings—meanings perhaps as numerous as the earthbound viewers of the televised events.

There are immediate meanings. For example, the highly successful mission of Apollo 11 means there will be more of the same. Apollo 12 is already scheduled to take off for a moon landing on Nov. 1. The landing site has been chosen, and there will be eight more moon landings after that, approximately four to six months apart. [Due to subsequent budget cuts, there actually were only six more moon missions and five landings. Apollo 13 did not land, due to an on board oxygen tank explosion that occurred the day before two of its crew members were to land on the moon.]

The next nine moon landings, according to space officials, are expected to show whether further lunar exploration would be feasible, and if so, whether moon bases might be both feasible and practical.

At first glance, the flight of Apollo 11 already has indicated the moon is a more interesting place than had been thought.

Meanwhile, beginning about 1972, the Apollo Applications Program—an Earth-orbiting workshop-laboratory program—is to start, using the basic Apollo vehicle and a modified upper stage of the Saturn launching vehicle.

Beyond that, no overriding space goals exist that are comparable to the one of landing men on the moon in the 1960s that was set by the late President John F. Kennedy in 1961. Just what such a goal might be is somewhat unclear now, but on frequently mentioned is a manned Mars mission for sometime in the 1980s. That trip would take at least 18 months, compared to the eight-day round trip of Apollo 11 and probably would involve eight to 12 astronauts, perhaps in two separate vehicles.

All eyes here focus on President Nixon, to see what recommendations he will make on this nation's future in space. The feeling here is that he is looking favorably upon space activities, especially since the triumphant journey of Apollo 11.

But in addition to meaning that man has now become a space explorer with an immediate future in space, the mission of Apollo 11—like a good book or play—stimulated thought, touched emotions, raised spirits and impressed the mind with lasting images.

It was a monument to an assassinated American president; it was the convergence of eight years of hard work and a sacrifice by more than 300,000 Americans—a convergence of efforts to a single point in time, Sunday, July 20, 1969 A.D.

It was a victory on several levels. The Russians were beaten to a manned lunar landing, despite what appeared to be their reminder, Luna 15, that they were still in the game.

It was a victory of human intelligence and achievement against an even more formidable antagonist—nature herself, who gave birth to that intelligence, but at the same time placed countless limits on the bodily forms harboring it.

There were emotional components, too. News reporters from around the world crowded into the National Aeronautics and Space Administration auditorium here last Sunday night to watch the television pictures of Armstrong's first step on the moon. They cheered when that step was taken; they rose to their feet with a standing ovation when the American flag was planted on the moon.

From around the Manned Spacecraft Center here, there were reports of men breaking into tears. And when Apollo 11 landed Thursday and Armstrong, Michael Collins and Aldrin were safely aboard the recovery ship, one or two top space officials found it difficult to speak.

"I've been trying to think of some moment in my life that compared to the moon landing," said one reporter. "The closest I can come to it was when my football team won the state championship—but that's not a very good comparison."

"That moment of exhilaration," said another, "cost each American $15 a year for the past eight years. That's $1.25 a month—probably less than the electric bill for one's daily television 'exhilarations.'"

"I had to pay a $20 bet to Walt Cunningham [an astronaut]," said another. "I bet him that we wouldn't land men on the moon by 1970. That's a bet I'm happy to pay off."

And there were unforgettable scenes. There was the shadowed leg of Armstrong moving slowly down the ladder of a spacecraft on an alien world a quarter of a million miles away; there were the kangaroo-like and ballet-like movements of Armstrong and Aldrin across a world that had, according to them, a stark beauty of its own.

But there were scenes here, too. At the moment of the moon landing, a poverty group broke into a signing demonstration at the foot of a full-scale mockup of the lunar module. They wanted the elimination of hunger and poverty to be the next American goal.

Throughout the eight-day journey of Apollo 11, many people here attempted to give expression to their thoughts. Some had to do with the speed of technological progress. This was the century that opened with the first successful airplane flight and, before it was three-fourths completed, saw human beings land on the moon.

This was also the century that saw men explore the last unknown continents on this planet.

The phrases, "end and beginning" and "new era for mankind" were heard frequently here and elsewhere. The moon landing was sometimes likened to the voyage of Columbus; sometimes to the first time aquatic life crawled onto dry land to begin a major chapter in the evolution of life.

To Dr. Wernher von Braun, the rocket scientist who has long dreamed of space travel, the Apollo 11 mission was not only equivalent to the beginning of dry-land evolution, but it also has virtually "assured mankind of immortality" because it demonstrated man's ability to travel to and live on other worlds.

Others here think the Apollo 11 mission is the beginning of expanded human vision. Ocean creatures cannot know of the existence of the stars and the vastness of the universe unless they leave the ocean, they argue. Similarly, humans' vision and consciousness will be expanded as they leave their "ocean" and explore worlds beyond.

It has been suggested that space travel may awaken new centers of our religious sense.

There are, of course, no ways of making firm evaluations of such ideas and speculations at the present. As Dr. Thomas O. Paine, NASA's administrator, said, "This success is something that has raised the spirits around the world, and it has caused us to pause and ponder its meaning, which only history, in the final analysis, will reveal to us."

But one or two things are clear. A new kind of hero and explorer has entered the human scene.

No longer is it sufficient for the modern hero and explorer to possess brute physical strength or skill alone; human endurance is no longer the only requisite.

Physical skill and endurance are, of course, still important, but something else has crept in—mental skill and mental endurance.

A lesson from Apollo 11 is that history's first moon explorers were men in top physical condition and also were intimately familiar with computers, mathematics, physics, astronomy and geology. They had to pay a price of mental discipline before a new world was opened up to them.

And the same was true for many of those more than 300,000 American workers who directly participated in the Apollo program.

That, perhaps, may be a commentary on the current evolution of man—it's taking a turn toward the mental, rather than continuing with the physical.

Therein may lie a more profound implication of the flight of Apollo 11.

Apollo 13—A Bird Cried...the Vigil Began
APRIL 19, 1970

HOUSTON—The scenes were like a strange mixture of an art film and an adventure movie. It was deep in the night. The American flag outside the Manned Spacecraft Center's administration building drooped; the wind had died.

Outside, among the buildings, all was quiet except for a bird— some think it was a Texas mockingbird—that emitted shrill, piercing notes. More than one person later commented on the bird; a physician wondered to himself whether it was hurt or wounded. In any case, he thought it strange that a bird should be crying, if that's what it was, at that hour.

At the main gate to the Manned Spacecraft Center, cars were racing in, slowing at the guard's post until recognized, then speeding on. The guard stopped some, made entries on a clipboard note pad, then waved them on.

Around the buildings, where the bird's cries pierced the night, an occasional figure could be seen running. They were news reporters.

Inside the news center, two sheets of paper were taped onto a counter top. There was a crowd around the two sheets of paper, focusing on some lines marked by a pen.

"SC [for spacecraft]: Okay, Houston. Hey, we've got a problem here."

"**CAPCOM [for capsule communications]: This is Houston. Say again, please.**"

"**SC: Houston, we've had a problem. We've had a main B bus interval.**"

The rushed stenographer listening to the tapes misunderstood. The word was supposed to be "undervolt," not "interval."

The marked lines continued:

"**SC: Okay—and we had a pretty large bang associated with the caution and warning there....**"

The space agency's public affairs officers looked grim. Questions were being fired at them, but they could not explain the meaning of those marked lines to the newsmen. There was anger, cursing.

Thus began on the night of April 13, the drama of Apollo 13, the spacecraft that took off for a mission to the moon on the 13th minute after the 13th hour on Mission Control clocks, in the 13th year of the space age. The strings of 13s, like the piercing whistles of the bird, became a conversation piece.

Thus, also, began a rescue mission that became a concern of nearly the whole world for the next 87 hours. It was a rescue mission that involved thousands of people around the nation, working as a unit. It was a rescue mission that involved the orchestration of mathematics, computers, technical devices, spacecraft simulators, many varieties of engineering, chemistry, physics, make-do inventiveness—and, most of all, human judgment, endurance and courage.

It was a blend that brought three men and a seriously disabled spacecraft around the moon—from a point 200,000 miles from Earth at the beginning of the crisis—and back to Earth into the rolling Pacific Ocean. That was the basic plot that welded the scenes together.

But that night—the night of the 13th—there was no happy ending in sight. In a rapid rush of events, the interchanges between astronauts James A. Lovell, Jr., Fred W. Haise, Jr. and John L. Swigert, Jr.—who was there because another astronaut was susceptible to German measles—and Mission Control became increasingly urgent. The voices of all were tight.

"**We are venting something out into space...it's a gas of some sort...I don't have any current now. Hey, it's off. They—they killed the bus completely now...It appears that we are losing the O2 [oxygen] flow...I've got a positive pitch rate and I can't stop it.**"

Such were the remarks, dispersed amid floods of technical talk about electrical circuits and spacecraft procedures, that came from the crew during ensuing minutes and quarter-hours. The voices may have sounded urgent, but they were never panicky.

And from Mission Control, amid its flood of technical words, numbers and acronyms:

"We're still trying to come up with some good ideas here for you...Okay, can you tell us anything about the venting? We'd like you to isolate your O2 surge tank...Stand by on those readouts...We are going to have to have you go through shutdown procedure on fuel cell one...We confirm that here, and the temperature also confirms it...There's no need to worry about that now."

Just listening to the conversation between spacecraft and Mission Control—to its pace, its rapidity, its tune—suggested something was very wrong, even if the content were not wholly comprehensible.

But then came exchanges that gave a chilling realization that something was not only very wrong, but grave.

"CAPCOM: It's slowly going to zero and we're starting to think about the LM lifeboat.'

"SC: Yes, that's something we're think about, too..."

And minutes later:

"CAPCOM: We figure we'll get about 15 minutes worth of power left in the command module. So we want you to start getting over in the LM and getting some power on that. We'd like you to start making your way over to the LM now."

The crisis struck near the end of a team of flight controllers' shift at Mission Control. The flight director on duty, Gene Kranz, realized his team was getting tired, too tired perhaps for the critical situation. The team coming on arrived an hour early. There was a quick briefing, and before the fresh team came in, Kranz announced, "Okay, all flight controllers, I'm handing over. I assume the majority of the new team guys are pretty much briefed and up to speed."

Elsewhere around the Manned Spacecraft Center—a huge campus of white, modern buildings set amidst a former Texas cattle pasture—the tempo of the movie-like scenes increased.

The hidden bird continued its piercing whistles.

Upper floor lights in the main administration building were on. Offices of the center's top officials and directors are on the upper floors.

There were phone calls to the astronauts' families; there were phone calls to Washington. Within hours, direct contact with the White House was established from Mission Control. The President was personally briefed on the situation by the space agency's administrator.

The vice president, who was to have visited the center the next day while the moon mission was underway, canceled. "He felt it would be inappropriate," a spokesman explained.

About two hours after the crisis began, a press conference was held with three of the moon program's leaders. It was held in a tiny auditorium made hot by television camera lights and a standing-room-only crowd of news reporters. A television cameraman shouted for newsmen to get out of his camera's way; several newsmen snarled back. The quarrel almost downed the opening words of the Manned Spacecraft Center's deputy director, Christopher C. Kraft, Jr.

"Well, I guess we should start out by saying that we have a serious problem in the command and service module. We appear to have some kind of accident with the—in the region of the fuel cells and the oxygen tanks. We have not tried too much to reconstruct the—what has happened because we're more concerned at the moment for getting the situation under control."

The director of flight operations discussed "options."

He said, "The minimum return-to-Earth time—this would be a total flight duration—would be about 133 hours; that would result in a landing in the Atlantic. That's one option we have. The second option would be go to the mid-Pacific line. That would take about 142 hours total flight duration...."

Then, the questions: "How much electrical lifetime do we have in the LM and how much oxygen lifetime? How long do we have?"

"Well, it depends upon how much we use it," came the answer.

How does the crew feel about the situation, and how do their wives feel?

"I would only guess that the crew feels that the situation is under control and they were in a serious condition and they knew they had a job to do, and I don't think that they stopped to consider what their personal feelings were at the moment. As far as the wives are concerned, I'm sure Deke Slayton [flight crew operations chief] has talked to them." That was Kraft answering the question.

With the breakup of the press conference, the late night mood slowly dispelled. The crew of Apollo 13 was settling down to the business of adapting to the lifeboat named Aquarius; the tasks of mission officials, scientists, engineers and technicians were coming into focus, and they all were beginning to carry them out.

The movie-like scenes faded into the subsequent days and nights, revived on the day of entry and splashdown.

The closing scene was that of three smiling astronauts, faces heavy with 5 o'clock shadows, walking across the recovery ship's deck while a band played:

"This is the Dawning of the Age of Aquarius...."

The Story of Apollo and 'Camelot'
DECEMBER 20, 1972

"Ask every person if he's heard the story,
and tell it strong and clear if he has not,
that once there was a fleeting wisp of glory
called Camelot."
From "Camelot," by Alan Jay Lerner and Frederick Loewe

HOUSTON—The manned lunar landing program was born more than 11 years ago with a pronouncement by a president, the late John F. Kennedy, who had a fondness for the then-current Lerner and Loewe musical, "Camelot," the legendary kingdom of King Arthur.

It was obviously for that reason that the Apollo 17 astronauts, the crew of the final mission in the Apollo moon landing program, named a prominent crater in the last landing site of the program Camelot.

The crater Camelot was a landmark crater, guiding Eugene A. Cernan and Dr. Harrison H. Schmitt, to their predetermined landing spot in a tiny mountain valley near the Sea of Serenity.

And seconds after they lifted off from the moon, after a highly successful scientific expedition on its surface, Schmitt noted the landmark a final time, "...And we're going, coming right over the top of Camelot."

Thus ended man's final visit to the moon for perhaps decades to come in a program that first took men to the moon only three and a half years ago. The glory of Apollo, as with the legendary kingdom Camelot, was therefore brief and fleeting, although its main product, knowledge of the moon and maybe of the evolution of the solar system, promises to unravel in the years to come.

But the active part of Apollo is over, and its story is told in different ways.

Some say it was necessary because man is by nature a curious creature, and it is as important for him to satisfy his exploratory urge as it is for him to have food, shelter and clothing.

Some say it was a monumental height of irrationality.

Some say it was nationalistic extravagance in a time of need and want in the very nation that launched it.

Some say it was the story of triumph over a cold war enemy, with the planting of six American flags on the Earth's only natural satellite.

And there are some who, like author Norman Mailer, have difficulty deciding whether landing men on the moon was the noblest expression of the century, or the "high mark of the fundamental insanity of the time."

There are certain documentable aspects to the story.

At one level, it was the story of how several hundred thousand people joined together to form a sort of collective consciousness to solve an exceedingly difficult problem.

The problem was to create a bubble, a module it came to be called, that would fully sustain three human beings as it drifted a quarter of a million miles in cold, hostile space to a celestial body equally hostile, and back to Earth again.

It was a problem of sending that bubble from a body that spins and moves, to another body that spins and moves, and to land its living cargo within hundreds of feet of a predetermined spot on that spinning, moving body. It was a problem of bringing the human cargo back to a given pinpoint on their spinning, moving home planet.

It was a problem of creating the machines, the devices, the communications to aid in the accomplishment of the task. Jules Verne had, in broad brush fashion, worked out some of the answers for a novel more than a hundred years before, but it took computers, radio and materials and concepts that he never dreamed of for the real thing.

In the end, the collective consciousness produced a rocket with more than five million functional parts that carefully and precisely controlled a violent explosion to launch the bubble from Earth.

In the end, the collective consciousness produced a bubble with more than two million functional parts, and still another with an additional two million working parts to land human beings on the moon.

An extra drop of solder on each electrical connection, or one thirty-second of an inch excess of electrical wire in the final combined products of rocket and modules would have added extra tons of prohibitive weight, making the whole thing unworkable.

But everything worked with the concert that the collective consciousness had planned and the fruits of the collective consciousness produced moments that lifted the spirits many on the home planet.

There was the Christmas flight of Apollo 8, ending on a positive note a year that saw two political assassinations in the United States.

There was the Apollo 11 moon landing, when Neil Armstrong and Edwin Aldrin became the first human beings to set foot on extraterrestrial soil. The world gave a collective cheer.

There was Apollo 13, when a collective consciousness of flight controllers on the ground quickly responded to effect a deep space rescue that has no parallel in this world.

There were Apollos 12, 14, 15, 16 and now 17 that were a progressive series of scientific expeditions, the results of which are leading to a growing familiarity with the once-mysterious moon on the part of the world's scientific consciousness.

And out of the Apollo program came a general awareness, imprinted on the collective consciousness of many by photographs and television images, of the Earth as a rich, colorful oasis in the black vastness of space.

The Earth has come to be viewed as a fragile bubble of life in space; somewhere out of it all came the term, "Spaceship Earth."

Thus, the collective consciousness that sent human beings out of this world has led to a turning inward. We have gone into space and seen our own planet in a renewed perspective.

Maybe ourselves, too, it is hoped by many.

The power of a united, collective consciousness may, in the end, be the story to tell strong and clear about the fleeting wisp of glory that was Apollo.

HISTORICAL (i.e. DATED, QUAINT)

Time present and time past
Are both perhaps present in time future,
And time future contained in time past.
T.S. ELIOT, *Four Quartets, Burnt Norton*

The first piece in this section was part of a staff-written series that attempted to sum up the decade of the 1960s and to look ahead into the 1970s. I, along with one or two of my colleagues, had a running battle with editors about ending decades a full year earlier than they technically concluded (the end of the 1960s came at the end of 1970, not 1969), but we always lost on the grounds that we reflect on the past decade when the numbers change. If the truth be told, I probably would have written the same thing at the end of 1970 that I wrote at the end of 1969.

This piece, flawed by occasional excesses of optimism and gloom, probably did capture some of the moods of the times. In revisiting it, I was struck by just how pivotal and vital the 1960s were to the science and technology that exists as we approach the new century (which, incidentally, begins January 1, 2001, not January 1, 2000). The 1960s were my first full decade of science writing, and they contained the seeds of most of the topics I was to cover in subsequent decades.

Also included in this section are two reports, written at the ends of the first and second weeks of a two-week, landmark trial that grew out of the first human heart transplant operation performed at the Medical College of Virginia. For the first time anywhere, a jury decided that brain death could be a criterion for declaring a person dead, thereby establishing a favorable legal climate for organ procurement for transplantation in this country. The plaintiff's lawyer in that case, incidentally, was L. Douglas Wilder, who became the nation's first black governor.

The '60s: A Decade for Science
NOVEMBER 2, 1969

The decade of the '60s was a complicated, paradoxical child of the time, at once precocious, fretful, noble, base, outgoing and darkly introspective.

It was the decade of lasers, emitters of pure and powerful beams of light; of DNA, the master molecule of life; of man in space, from orbits around the Earth to footprints on the moon.

It was the decade of transplants, when people were living with other people's kidneys, hearts and other tissue within them. It was the decade that brought plastics and electronics into human bodies— Dacron blood vessel grafts, Silastic heart valves and compact pacemakers to spark failing hearts into rhythmic beating.

It was a decade that saw polio dwindle to almost nothing in public health lists, and measles declined.

It was a decade of closeup pictures of the planet Mars, of moon rocks and soil in Earthbound laboratories. It was a decade in which the "noise" of the universe being born was detected.

It was a decade in which people began seeing a universe stranger than human imagination could conjure; it was the decade of quasars and pulsars, strange objects in space that strained the physics and astrophysics of the moment for explanations.

It also was a decade in which science-based technology scattered its fruits among average citizens. Tape recorders fell within the means of the average person; walkie-talkies became playthings for children. Transistor radios about the size of cigarette packages became ubiquitous.

The promise and potential of television solidified within the world culture, with communications satellites helping knit continents, and major events within them, together on a real time basis. The dimension of color was added to many home screens.

In a broader sense, the decade of the '60s was a foundation decade, a footing for much of the science and technology, with their promises and woes, of the '70s.

The decade of the '60s was the decade in which computers broke out of their shells of limited, selected usage and became entrenched in the American way of life.

They began running machinery. They began keeping records, issuing paychecks, handling orders and serving routinely as abrasively accurate know-it-alls on election nights.

And it was a decade of scientific and technological events that Virginia helped shape. The nation's manned space flight program was born and nurtured at the space agency's Langley Research Center; numerous contributions to the moon landing program—including the basic flight plan—came from there.

Fundamental strata in the field of organ transplantation were laid down by medical scientists at the Medical College of Virginia, where approximately 170 kidney transplants were performed during the '60s. The world's longest surviving heart transplant recipient is a man whose procedure was done at MCV.

It was also a decade that saw an increasing interest in some new, or formerly obscure, disciplines of science, such as oceanography, ecology, ethology (the study of animal behavior with implications for understanding human behavior), immunology, biometry and bionics (applying biological processes and functions to the design and development of man-made engineering systems).

It was a decade in which science and technology added many new words to the language, or gave new contexts to some old words, such as "transplant" (medical) and "rendezvous" (space).

It was also a decade in which science and technology became socially relevant in numerous direct and indirect ways. Automation became an issue in labor-management disputes. Current views, from biology and genetics, on the importance of the environment in shaping individuals with given genetic endowments became an underlying philosophy in social projects for the underprivileged.

It can even be argued that the "God-is-dead" discussions of the latter 1960s turned largely around scientific and technological issues. It was a decade in which people were struck with the realization of their own power and potential, and of their responsibility for their decisions.

And a rising interest in the occult—particularly a mass popularity of astrology—has been claimed by many observers to be a reaction to the dominance achieved by science and technology during this decade.

Moral and ethical issues became topical during the 1960s largely because of medical science's growing power over death. Machines and drugs became increasingly capable of stalling off "natural" death, thus raising questions of how far heroic efforts should go to preserve life.

The era of heart transplantation brought to the fore questions about definitions of death—questions based on modern knowledge that cessation of heartbeat, the traditional basis for definitions of death, is an entirely inadequate criterion.

Chemicals, too, became socially relevant during the 1960s. There was thalidomide, the drug that caused malformed babies because their mothers took the agent to combat morning sickness. Although the problem arose in Europe, it had far reaching repercussions in this country, raising among other things, legal, ethical and philosophical questions about drug testing and safety.

The birth control pill became especially relevant during the '60s. The Pill has been credited with liberating the American woman, even with fostering a "new morality" and a "sexual revolution."

It was also the age of "Better Kicks Through Chemistry," as one drug official dubbed the "Now Generation's" preoccupation with drugs as a means of bearing, distorting or escaping from a reality created by a scientific and technologically based society.

Chemistry became socially relevant in still other ways during the 1960s. On the one hand, there were new and better drugs, better and new synthetic materials.

Clothing, draperies, toys, upholstered furniture, shoes, disposable coffee cups, picnic ice chests and thousands of other daily items poured out of the chemical engineer's vats rather than from nature's resources.

The creeping prevalence of synthetic fibers swept over men, women, children and their possessions to an astonishing degree during the 1960s. With his vinyl shoes, his Dacron socks, his Dacron-polyester slacks, his acrylic shirt and his Dacron-blend knit sport jacket, the well-dressed man of the '60s stepped farther away from nature—her wools, cottons and leathers.

The creeping prevalence of chemicals in the environment—in food, in the air, in all manner of wildlife—also became of concern during the '60s. The price of chemical pesticides' benefits was realized during tis decade. Even the bodies of penguins in the isolated Antarctic—thousands of miles from the nearest pesticide sprayings—were found to contain DDT.

And the price of technology-based affluence was realized, too, as air and water pollution became major issues. The seriousness of the problem was talked about, and quietly manifested by foreboding scenes around the nation.

"The air and water grow heavier with the debris of our spectacular civilization," warned Lyndon B. Johnson in a conservation message to Congress in early 1967.

And the scenes included a dying Lake Erie, where the fish are almost gone and swimming is permitted only occasionally.

Along the entire Eastern Seaboard, from Boston to Washington, agricultural crops become glazed, blotchy and discolored because of the smog and its toxic products. The Eastern Seaboard, like southern California, now suffers agricultural losses each year because of the smog.

Breathing New York City's air for one day is equivalent to smoking two packs of cigarettes, said one report on air pollution there. And several urban rivers are so thick with flammable wastes they have been deemed fire hazards.

This was also the decade that the world's population explosion—largely a result of reduced infant mortality—became audible, and when the doctrine of a 19th century English clergyman, Thomas Robert Malthus, was debated. Malthus had said, in essence, that an unchecked population always outgrows its food supply.

By the end of the '60s, the world's population had reached 3.7 billion people, and more than one-half of that number were said to be malnourished, hungry or starving.

In short, the decade was one in which elements of science and technology became inextricably mixed with social, legal, moral, ethical and political problems. It was a decade in which the liabilities as well as the assets of science and technology were not only being recognized, but became social, legal, moral, ethical and political issues as well.

It was also a decade of great historical significance for science and technology, filled with events and developments that touched the lives of present individuals and which will undoubtedly affect the lives of generations to come.

The '60s and Cape Kennedy were for space exploration what Kitty Hawk and 1903 were for powered flight in the air. The unmanned exploration of the space around the Earth, of Venus, and of Mars, not to mention weather and communications satellites, marked the beginning of the extruding of man's senses and consciousness from his home planet.

And Apollo 11, the moon landing expedition of Neil A. Armstrong, Edwin E. Aldrin, Jr. and Michael Collins, heralded an era in which man himself will be purposely drifting, like an intelligent spore, through space.

Of all the technological developments, the journey of Apollo 11 is destined to become the most popularly remembered, even though there were other events that will be equally, if not more, significant for humanity's future.

Apollo 11 was a national goal achieved; it was a tip to America's scientific and technological iceberg; it was the television spectacular of the decade.

It was generally hailed as a positive and noble feat around the Earth, with most of the civilized world being united during those hours of July 20, 1969, cheering for two human beings who made it onto another world.

That, perhaps more than any other single event of the 1960s, dramatized how far humans had separated themselves from nature.

In fact, it was captured in one photograph—a photograph taken by Neil Armstrong of Edwin Aldrin standing on the blue-gray lunar surface, in a small crater undisturbed for millions of years, almost a black sky, dressed in unearthly attire of artificial fibers and materials. He wasn't made for this world but there he was anyway.

But even that event was not cleanly isolated from the rough and tumble of terrestrial life. The moon trip, said many, was of no earthly use. It was a matter of engineering, not science, said others. It was not conceived in the spirit of more noble human aspirations, but amid earthly politics, said still others, noting that the moon goal was motivated by what Russia was doing in space rather than by human curiosity about the universe.

And the moon trip became a symbol of the wide disparity between human technical and scientific skills and people's abilities to control themselves. "We can send men to the moon, but we can't stop wars"; "We send men to the moon, but we haven't conquered poverty or discrimination or pollution...."

Thus did the moon goal of the '60s raise questions about values and priorities for an affluent nation.

But while space exploration made the headlines, the nucleus for another revolution, with perhaps greater implications for the future, began taking shape during the '60s. This involved a deepening insight into the fundamental mechanisms of life, especially the DNA molecule, which lies at the heart of every living thing, from amoebae and oaks to gnats and human beings.

During the 1960s, scientists learned the nature of DNA's chemical code. They synthesized a relatively simple active virus; they made a human protein—insulin—in the laboratory for the first time.

The dawning of the day of "biological engineering"—of correcting mistakes in DNA's code to repair chemical and other defects in humans—perhaps broke during the '60s. The awesome capability of directing evolution became something more than a vague possibility during this decade.

Some biologists were already urging that steps be taken to consider the moral, ethical and philosophical implications of what one popular writer termed the "biological time-bomb," and to develop guidelines for such research now—before the bomb begins exploding.

So ends the decade of the '60s—a decade of unprecedented scientific and technological achievement; a foundation decade for the future, and a decade in which science and technology became more involved than ever before with issues.

Two Concepts of Death Converge in Trial Here
MAY 21, 1972

Two concepts of death were converging on a collision course in a Richmond courtroom last week, and the outcome could have significant implications for both the profession of law and the profession of medicine.

The convergence of the two concepts developed, first as an underlying issue, then a dominant one, in a trial in Law and Equity Court stemming from a heart transplant operation performed at the Medical College of Virginia in 1968. The brother of the donor in that operation, Bruce O. Tucker, is suing various MCV transplant surgeons for wrongful death.

Mr. Tucker had suffered severe head injuries in an accidental fall on a Friday evening in May 1968, and was admitted to MCV. An operation was performed on Mr. Tucker later that night.

The following morning, he was placed on a mechanical respirator, because, several physicians testified, he was no longer able to breathe on his own.

A neurologist who examined Mr. Tucker within two hours after his being placed on the mechanical respirator testified that from the standpoint of brain function, Mr. Tucker was dead and had been for some time prior to the examination.

Not only dead were the brain's higher centers—those involved with the complex mixture of functions popularly lumped together under the term "personality"—but so were the more primitive functions of the brain stem, such as control of respiration, the neurologist testified.

Thus, without the mechanical respirator, Mr. Tucker could not breathe on his own, the neurologist said. But with the mechanical device breathing for him, certain functions within his body were able to continue. The heart, for example, continued beating. Thus, there was a pulse; there was circulation; there was blood pressure; there was body temperature. Many cells, tissues and certain organs in his body, therefore, could be considered alive. Skin, whiskers and fingernails would continue growing.

At the instigation of the plaintiff's attorney, Lawrence Douglas Wilder, an anesthesiologist read in court last week his records of Mr. Tucker's vital signs, which include respiration, pulse, blood pressure and temperature, for varying time intervals before and after he was officially pronounced dead at 3:35 p.m.

For example, *"4 p.m., temperature, 93 degrees; blood pressure, 90; pulse, 115; respiration (provided artificially) 20.*

"4:15 p.m., temperature, 92.5; pulse, 95; blood pressure, 70; respiration, 20.

"4:30 p.m., temperature, 92.5; pulse, 100; respiration, 20; blood pressure, 80.

"...4:32 p.m., heart out."

Therein lies the issue concerning the two concepts of death.

There's the traditional concept, bound with various legal rulings and, perhaps, much popular opinion, that says death occurs when breathing stops and the heart stops.

That concept, especially as it involves cessation of the heart, is intimately linked to the case developed by Wilder. Removing Mr. Tucker's heart was really what killed him, Wilder has contended, and eliminated all further possibilities of resuscitation. The question is not so much when death occurred, but when life ceased, he said early in the trial.

There are certain legal guidelines supporting the more traditional concept of death, including the definition in *Black's Law Dictionary*.

Death, says that dictionary, is "the cessation of life; the ceasing to exist; defined by physicians as a total stoppage of the circulation of the blood, and a cessation of the animal and vital functions consequent thereon, such as respiration, pulsation, etc."

The other concept of death, as propounded by the defendants' lawyer, Jack B. Russell, is intimately linked to brain death. Once the brain is dead, and biological death of the remainder of the body's tissues and organs proceeds at varying rates, he indicated.

Thus, death does not occur at a precise instant, but is a continuing process, occurring at different rates in different parts of the body, according to that concept, with the death of the brain being the key indicator, since the personality and other characteristics of the person are centered in the brain.

People may be resuscitated if the heart and respiration halt for a period of time, and be functional human beings again after such resuscitation, another defense lawyer, Asst. Attorney General Theodore J. Markow suggest during Friday's proceedings.

Black's Dictionary definition of death, he further contended, "is not a definition," but only lists "signs or evidence that death has occurred." That kind of evidence had been developed over the years by the medical profession, but now, in light of newer medical evidence, the definition "doesn't go far enough."

In the specific case of Mr. Tucker, the neurologist had testified that his examination showed a man who was irreversibly dead from a neurological standpoint, incapable of sustaining breathing on his own.

The neurologist also testified that circulation of blood, which provides oxygen to keep cells alive, had ceased to the brain; swelling and fluid accumulation within the skull and brain had shut off such circulation, even though the heart was pumping blood (because of the mechanical respirator's action) to other parts of the body.

The neurologist also testified that he had "never heard or seen or read of any case that has recovered with the evidence that we had in this case."

There has been a growing trend within many segments of the medical profession, especially with the advent of the era of transplantation, to base a concept of death on the brain rather than on heartbeat and respiration alone.

In 1968, for example, a Harvard University panel composed of physicians, lawyers, theologians and philosophers recommended criteria for brain death that were published in the *Journal of the American Medical Association*.

Those criteria for judging a patient dead, and justifying the shut-off of machines supporting the patient, were: No spontaneous respiration; no superficial or deep reflexes; no responsiveness to stimuli, including deep pain, and no sign of brain activity on an electroencephalograph (EEG).

To cover those cases in which exposure to cold or an overdose of drugs may have been responsible for suppressed brain activity, the panel recommended retesting of the patient 24 hours later.

Other approaches to diagnosing brain death have also been devised by various groups of physicians, in which the clinical judgment of the attending physicians is emphasized.

Testimony at the trial during the past week indicated that total lack of brain activity in an EEG recording; no spontaneous breathing; absence of various reflexes, and lack of responsiveness to painful stimuli were among factors taken into account in judging Mr. Tucker dead from a neurological standpoint.

Besides questions concerning the head-on course of the two concepts of death, questions have also emerged during the trial, which is to resume Tuesday morning, about whether judgments about death should be left in the hands of physicians or whether they should become legal matters.

"If the court is going to adopt Black's definition or one similar," said Russell Friday in arguing a motion for dismissal of the case against the MCV surgeons, "then every single donor, without regard to whether there was or was not consent, had his life terminated. A person cannot give consent to an unlawful act. Every transplant was illegal."

If it became a legal matter, he wondered whether a physician would have to seek a court's permission before turning off a mechanical respirator on any patient.

Wilder argued on the other hand, "Unless we have a legal definition of death, we will be indeed in hard times."

Medical, Legal Paths Finally Crossed
MAY 29, 1972

Certain powers that medicine began acquiring about a decade ago were almost bound to cross legal paths sometime, some place, somehow, and they finally did.

The time was the past two weeks. The place was Richmond's Law and Equity Court. The occasion was a trial that resulted from a human heart transplant operation performed at the Medical College of Virginia four years ago.

In a word, medical science had created a plug that could be pulled, thereby allowing traditional and legal signs of life (pulse, temperature, blood pressure) to cease. At the other end of the plug is a device called a "positive pressure respirator," a machine that performs the rhythmic breathing functions for a patient.

Also approximately within the past decade or so, physicians learned how to restart hearts that had stopped beating, and that capability was soon found to be a mixed blessing.

The heart, it must be remembered, pumps oxygen-laden blood through the body, providing cells (which make up tissues and organs) with the oxygen they need to remain viable. Without oxygen, they die, and the cells of some tissues and organs die quicker than others.

The cells that make up the brain are the most sensitive to oxygen deprivation. If deprived of a flow of blood for four or five minutes, brain cells begin dying, and, unlike many other cells, brain cells do not reproduce. Thus, there is no replacing of dead brain cells; there is no healing of the brain once its tissues are damaged.

So, those two medical developments within the past decade—mechanical respirators and means and knowledge of resuscitating stopped hearts—began creating problems as well as solving many. The problems that were created suddenly brought once abstract issues of life and death into the realms of routine medical practice.

A person's brain tissue may have been totally damaged as a result of a stroke or injury, but with the mechanical respirator, oxygen is supplied to the lungs, to the blood, to the heart muscle, allowing it to continue its pumping. There is a pulse because the heart is beating; there is blood pressure because blood is being pumped; there is temperature because many of the body's cells are still alive and active, and metabolism is continuing.

Or a heart attack may have led to heart stoppage. Resuscitative measures may have gotten it going again, but after the critical time limit for the brain's demands for oxygen had passed. A resuscitated individual with no brain function thus may be the result. Again, a mechanical respirator could, in principle, support the patient's vital signs of pulse, blood pressure, temperature.

Intravenous feeding could further support such patients by providing the things the cells need, beyond oxygen, to maintain themselves.

As one neurosurgeon testified during the recent trial here, "We found ourselves frequently [during the early 1960s] with a resuscitated patient whose brain was dead, essentially a cadaver whose heart was beating. Doctors now faced a problem of what to do...We were frequently faced with the problem of hospital beds filled by patients whose brains were dead and whose organs were being maintained."

By legal precedent, by legal definitions, however, death is a state devoid of vital signs, like blood pressure, pulse and temperature. Those patients had those. But their brains, medically the centers of their personalities, were totally and irreversibly dead.

Thus, a dilemma: Should you pull the plug? When should you pull the plug? Who should pull the plug?

Medical progress was beginning to cross legal paths.

Increasingly during the 1960s, various medical centers and medical groups began adopting irreversible and total brain death as being synonymous with death itself.

In 1968, a 13-member ad hoc committee of the Harvard Medical School, a committee compose of physicians, lawyers, theologians and philosophers put it together in a now famous but now somewhat dated report that began:

"Our primary purpose is to define irreversible coma as a new criterion for death."

It went on, "There are two reasons why there is a need for a definition: (1) Improvements in resuscitative and supportive measures have led to increased efforts to save those who are desperately injured. Sometimes these efforts have only partial success so that the result is an individual whose heart continues to beat but whose brain is irreversibly damaged. The burden is great on patients who suffer permanent loss of intellect, on their families, on the hospitals, and on those in need of hospital beds already occupied by these comatose patients. (2) Obsolete criteria for the definition of death can lead to controversy in obtaining organs for transplantation."

The transplant issue was where the course of medical progress directly intersected with the legal path. In May 1968, a Richmond man sustained a severe head injury that, according to trail testimony here, rapidly led to irreversible and total brain death. He was placed on a respirator at MCV when his breathing stopped. His vital signs persisted, however, as long as he was on the mechanical respirator.

His brain suffered devastating damage, but the remainder of his body tissue and organs were intact, which made him a prime candidate to be a donor of certain organs for transplantation in patients whose own heart or kidneys had failed or were failing. Keeping him on the respirator would help maintain the viability of those organs, for a number of hours, anyway.

Transplant surgeons in medical centers around the country and around the world had learned this some years before 1968, and rather routinely the mechanical respirator was used to preserve a donor's tissue and organs until close to the time of transplantation.

In the particular case here, a family member appeared at the hospital after the man's heart and kidneys had been removed and the heart was transplanted into another man's body. The surgeons said no relatives of the donor were known, or could be located before or after the donor had died, in the brain-death sense of the word, so he fell under Virginia's unclaimed body statute. That statute has provisions for the use of unclaimed bodies "for the advancement of medical science," it was contended in the court case that ultimately resulted.

In that case, the donor's brother brought charges of wrongful death against the surgeons. He claimed that his brother was not legally dead until his heart was removed, because his brother's vital signs continued to the moment the heart was excised.

Thus, the capabilities of medicine crossed the path of law. The newer definition of death adopted by medicine as a result of problems created by medical science was directly in the path of the legal definition.

So, during the past two weeks, the confrontation developed and finally stood as a stark, major issue in Richmond's Law and Equity Courtroom, about a block away from the Medical College of Virginia, where the elements of the issue began four years ago.

A jury of laymen ruled in favor of the surgeons, which meant support of the medical judgment of death and support of the new medical definition of death.

Meanwhile, the law's path may become more aligned with medicine's. There already are straws in the wind. The Kansas legislature in 1970 enacted a definition of death that is essentially the medical definition that includes brain death. It also stresses the "opinion of a physician" in the determination of death. The Maryland legislature enacted a similar definition this year.

By the end of last week, some observers were speculating that the outcome of the unprecedented trial here, and the attention directed to it nationally, may lead other state legislatures, lawyers, courts and the public itself to do some serious soul-searing about the once abstract, now practical question:

What is death?

Second Thoughts Followed Joy
JULY 20, 1995

Many of them were jubilant the morning the bomb went off.

Physicist Richard Feynman sat at the end of a jeep and played the bongo drums. Some shouted. Some danced.

"Everybody had parties; we all ran around," Feynman recalled 40 years later in his book, *Surely You're Joking, Mr. Feynman.*

"Everybody" was a large group of scientists, mathematicians and engineers at Los Alamos, N.M., who developed the atomic bomb. On July 16, 1945, at 5:30 a.m., many of them observed the explosion of the world's first nuclear weapon in an isolated part of the desert, in a test that was given the code name Trinity.

For the most part, the scientists were elated at the success of their several years of intense, esoteric work. According to accounts by Feynman and others, there was considerable jubilation at the observation site, on the buses that carried them back to the Los Alamos complex and at the complex itself.

But then came the second thoughts, earlier for some than for others.

J. Robert Oppenheimer, the director of the atomic bomb project at Los Alamos, recalled thinking of a verse in the *Bhagavad-Gita*: "Now I am become Death, the destroyer of worlds," when the awesome blast occurred.

The test director, Kenneth Bainbridge, congratulated everyone, then told Oppenheimer, "Now we are all sons of bitches."

I.I. Rabi later described his initial elation at seeing the fireball. But then he felt a chill, not from the morning cold, he said, but from thoughts of his wooden home, his laboratory, of the millions of people who lived back in the Northeast. He thought of all that in the context of the destructive power of nature that the scientists had unleashed.

Feynman recalled that his friend and colleague, Bob Wilson, was moping amid the jubilant scientists at the Trinity observation site that morning. "It's a terrible thing that we made," Wilson told Feynman.

Some started having misgivings well before Trinity. One of the prime movers behind the project was physicist Leo Szilard, who began arguing before the Trinity test that the bomb should be outlawed and never used.

Edward Teller, who later was instrumental in developing the more powerful hydrogen bomb, disagreed but noted in a response to Szilard that he had no hope of clearing his conscience. "The things we are working on are so terrible that no amount of protesting or fiddling with politics will save our souls," Teller said in the opening of his letter.

Feynman's own dark reactions to the bomb descended on him over the following months, after he left Los Alamos and returned to Cornell University to teach.

He told of sitting in a restaurant in New York one day, looking out at the buildings around and suddenly thinking of the bomb that was dropped on Hiroshima, and of the destruction it would cause if it were dropped on Manhattan. All the buildings he saw would be smashed.

He said that when he saw people building a bridge or a road, he thought they were crazy because it was all so useless.

But fortunately, Feynman said in his book, he turned out to be wrong about the uselessness of building bridges, "and I'm glad those other people had the sense to go ahead."

The scientists who worked on the bomb project did so because they believed in what they were doing. They thought it was necessary, since they had good reason to believe that scientists in Nazi Germany were also working on an atomic bomb.

But success meant that the American scientists became agents for unleashing unprecedented destructive power and terror into the world. Until then, the pursuit of science had been a relatively benign endeavor.

The Trinity test changed that in an instant, forever.

Scientists Can Be Humanly Wrong,Too
FEBRUARY 7, 1982

Science may be a good way of establishing certain truths about the physical world. But don't always rely on scientists. In many ways, they can be as humanly wrong and emotional as the rest of us.

Despite popular impressions, scientists are not always open minded, unbiased or humble before the facts. Nor are their judgments necessarily any more unerring than those made by people in other disciplines.

Some examples are classics.

In the 1830s, for instance, a British scientist named Dr. Dionysius Lardner proclaimed that a steamship would be unable to cross the Atlantic Ocean because it would need more coal than it could carry. A few years later, the first steamship crossed the Atlantic Ocean.

Dr. Lardner, professor of natural philosophy and astronomy at University College at London, is also said to have stated that human beings could not withstand travel at high speeds, like speeds of 100 miles an hour or greater, because their breath would be sucked out of their lungs.

And there's the famous pronouncement of the prominent American astronomer, Simon Newcomb, about the possibilities of a "flying machine." In an article published two months before the Wright brothers made history's first heavier-than-air airplane flight, Newcomb said that the fact that no combination of known forces, substances or machinery could yield a flying machine seemed to him to be as certain as any physical fact could be.

In some cases, entire bodies of scientists have turned out to be wrong. In 18th century France, for example, farmers would occasionally bring rocks to the French Academy of Science, claiming that the objects fell from the sky. The academy refused to believe the farmers, and dismissed them as cranks.

Thomas Jefferson, lauded for his scientific perception by many historians, also refused to believe that rocks (meteorites) could come from the sky. When two professors in Connecticut reported such an event, Jefferson is reported to have said he would rather believe that professors would lie than that stones would fall from heaven.

From more recent history, Thomas Edison is said to have believed that television would not amount to anything; nuclear power was proclaimed to be an impossible dream by some prominent scientists in the 1930s, and before the astronauts' landing on the moon, there were astronomers who predicted that the landing craft would sink in a sea of fine dust.

Such examples do not necessarily represent the usual course of science. Some of them are goofs that are remembered partly because we think they are funny, because we are in a position to know better, thanks to subsequent scientific progress and knowledge that we have the advantage of and they didn't.

But they do illustrate a larger point, namely, scientists can be as fallible as anyone else. And such a point is worth making. The human fallibility of scientists often tends to be overlooked; we often tend to take the pronouncements of scientists as being the definitive truth. We often tend to view scientists with awe, perhaps because we think they understand things that we do not, or cannot, understand.

Therefore, we may tend to consider scientists as being different, living lives untouched by the petty, biased beliefs that the rest of us carry around, using objective, unbiased data as the basis for their work and actions, and possessing a degree of intelligence that the rest of use don't have. As a result of such beliefs, we may tend to take a scientist's word for something more seriously than we take the word of anybody else.

What we too often don't realize is that scientists are humans, too. They, too, can be wrong. In a number of cases, it's subsequent findings that prove them to be wrong. That's as it should be; that's science at work. Hypotheses are made, even outlandish ones sometimes. Then, tests and observations are made to confirm or deny them.

But there also are instances in which their very humanness is a factor. Scientists can turn out to be wrong because of their biases. They can turn out to be wrong because something doesn't fit the established view.

A few years ago, an experimental psychologist, Dr. Michael Mahoney, sent two versions of a scientific paper to 75 journal reviewers. A journal reviewer is a scientist to whom journal editors submit manuscripts for his or her judgment on whether the report is worthy of publication. Reviewers are, in a sense, the gatekeepers of scientific "truths," with considerable power to say what shall pass and what shall not.

One version of the paper that Mahoney sent to reviewers supported the common wisdom of the day while the other version refuted it. He reported a strong bias on the part of the reviewers toward accepting the version that favored the widely believed hypotheses of the day.

(Some observers question the effectiveness of the reviewer-gate-keeper system. If a manuscript is rejected by a top quality journal, those observers say, it may be resubmitted to journals with lesser standards and is likely to be published eventually anyway.)

There also are instances in which scientists have not been particularly gracious or tolerant of views they considered nonscientific or unorthodox. In the 1960s, for example, the nation's leading science journal, Science, refused to publish a letter from a well-known astronomer, who for 20 years was the Air Force's consultant on UFO investigations, that suggested a tiny percentage of UFO reports may merit further scientific study.

An eminent American scientist led an effort in the 1950s to have a publisher stop publication of Immanuel Velikovsky's *Worlds in Collision*, a book containing some farfetched notions about the origin of Venus. More recently, some well-known scientists attempted to have the American Association for the Advancement of Science do away with its section on parapsychology.

The fact that scientists can act like ordinary people, rather than at the high planes of objective "truth," is further reflected by their emotional behavior at times. During debates in the late 1970s on controls of recombinant DNA experiments, there were instances of dirty-word name-calling from scientists on one side of the issue for those on the other side.

That issue became so charged, in fact, that at least one untenured Harvard biologist withdrew her participation in the debate for fear that her career would be harmed. A lot of others also dropped out, probably for the same reason, that biologist is quoted as saying in Nicholas Wade's book on the recombinant DNA controversy, *The Ultimate Experiment.*

Quibbling and name-calling do, indeed, occur in the arena of science. One New York science reporter noted that "an enormous amount of passion, vituperation and outright dirty-name-calling" arose over the rather esoteric issue of how closely one flightless bird, the kiwi, is related to another, the casoway.

The assumption that scientists are different from the rest of us is wrong, that reporter observed. They do not approach all problems in a logical, scientific way, he said, adding, "In fact, scientists are pretty much like us. Some of them are ascetic saints, some are drunken lechers, some do things just to get their name in the papers."

The human side of science is also replete with examples of politics, grandstanding and knee-jerk reactions. A young woman graduate student was in the forefront of one of the most significant findings about the brain in decades, a finding that led to the insight that the brain makes its own morphine-like compounds, but it was her male adviser and others who shared a major prize for the discovery.

Nobel Laureate James Watson described in a popular book, *The Double Helix*, the intense, sometimes ungentlemanly quest for determining the structure of the master molecule of life, DNA, because of the sure honor—like the Nobel Prize—that was to come for such a discovery. He, Francis Crick and others at Cambridge University were in a race with Dr. Linus Pauling to discover DNA's structure first.

And Dr. Pauling, meanwhile, was roundly denounced for speaking outside his field of expertise—indeed, there were even rumors that he was becoming senile—when he suggest that vitamin C should be studied for its effectiveness in preventing and treating the common cold. Such a claim was against the established dogma, and to a considerable degree, it still is, although a number of studies of vitamin C were prompted by Dr. Pauling's original book.

And just as in our daily lives there are generally acceptable and generally unacceptable behavior, there is in science certain topics that are comfortably acceptable and areas in which young researchers may do well to avoid if they wish to keep their credibility with their mainstream colleagues.

Parapsychology, UFOs, and vitamin therapy for mental illness (another Linus Pauling interest) are not exactly mainstream fields. The issue of genetic differences and genetic determinants of aberrant behavior is a sensitive social-political topic.

As far as intelligence is concerned, the public's intuitive notion that scientists are highly intelligent is probably valid. Mahoney, the psychologist, says that as a group, scientists are probably in the upper 5 percent of the population in terms of intelligence.

But whether intelligence is related to performance as a scientist is another question. In a book titled *Scientist as Subject: The Psychological Imperative*, Mahoney says the relationship is, at best, a hazy one. He says that the academic records of famous scientists are no better than those of their less eminent colleagues. He also says there is no correlation between IQ and the number of or quality of published scientific papers.

Mahoney also said he tends to believe that general and specific types of intelligence have little to do with high achievement in science.

Further, he questions the notion that the fraternity of science has exclusive entry standards. Mahoney claims that criteria for admission to graduate school are poor predictors of success in graduate school and after. He cited one analysis of admissions criteria to Pennsylvania State University that suggested that the length of recommendation letters was a significant predictor of acceptance. The analysis also found that successful attainment of a Ph.D. degree was strongly correlated with the first digit of one's Zip Code.

The fact that science has its human side is no surprise to those who are scientists, nor to those who follow it closely. Books ave been written on the subject; the "News and Comment" section of the journal *Science* regularly chronicles directly and indirectly such a face to science.

And there is considerable evidence that many scientists themselves do not believe the storybook images of them as dispassionate, objective seekers of truth. The ideal of science contains such attributes, and as a whole, the general drift of scientists may be toward that ideal.

In time, after all, the French Academy and scientists in general did acknowledge the fact that rocks indeed could fall from the sky.

LIFE—SWEET MYSTERIES OF

Now, my own suspicion is that the universe is not only queerer than
we suppose, but queerer than we can suppose.
J.B.S. HALDANE, *Possible Worlds*

Stars, black holes and the cosmos itself all have their mysteries, but
none are more intriguing than those manifested by the seemingly com-
monplace living things around us. How—or why—do affectionate touch-
es slow the buildup of cholesterol-laden debris within a rabbit's heart
arteries? What's the survival advantage of the starling's remarkable abil-
ity to mimic sounds, including human sounds? And where are the roots of
the deep organizing principles that influence flocks of birds, swarms of
gnats, schools of fish and collections of human cells?

Such mysteries repeatedly captured my fancy as column topics,
even though I could usually go no further than reporting their existence.
Someday, I'm sure, many of them will be cleared up by scientists, and the
answers will undoubtedly be as wondrous as the mysteries were.

Effects of Tender, Loving Lab Care
JANUARY 9, 1984

Several years ago, some researchers at Ohio State University
attempted to create a group of rabbits with atherosclerosis, by feeding
them a high cholesterol diet. Atherosclerosis, the process that clogs vital
arteries with fatty material, is the cause of many heart attacks, strokes and
angina in people.

But the researchers' efforts to establish such a group of rabbits for
their study were unsuccessful, and for a while, they couldn't figure out why.

The reason, it turned out, lay with how the animals were handled
by the experimenters. Affection, it seems, can significantly retard the
clogging of arteries, in rabbits at least.

More recently, researchers at Virginia Polytechnic Institute and State University also found positive effects of caring attention—on chickens. And in the mid-1970s, cancer researchers in Seattle found that an environment more respectful of experimental mice than the typical laboratory animal quarters resulted in fewer cases of cancer among cancer-prone mice.

It is best, according to a prevailing rule in the life sciences, to be cautious about extrapolating findings from animals to people, but it is tempting to speculate that the rabbits, chickens and mice are responding to a universal principle. That principle seems to be that biological systems operate best when their contacts with their surroundings are pleasant.

There is a considerable body of research indicating the bad effects of unpleasant environments, such as high levels of noise or unexpected shocks, on laboratory animals. Now, a small body of research on the effects of positive environments seems to be accumulating.

In the case of the Ohio State studies, reported in *Science* in 1980, it was found that rabbits fed high cholesterol diets and that were petted, handled and talked to daily by an experimenter had far fewer signs (60 percent fewer) of atherosclerosis in their arteries than did rabbits fed the same diet but essentially ignored by the experimenters.

The phenomenon came to their attention when they were unable to produce a sufficient degree of atherosclerosis in a group of animals despite the high levels of cholesterol in their diets. It turned out that a woman on the team had been visiting the rabbits several times a day to talk to and cuddle them.

The researchers, who originally intended to study the effects of certain drugs on high cholesterol diets, repeated the tender-loving-care experiment two additional times, each with similar outcomes, before making their report in *Science*.

Five years earlier, the same journal published a report by Dr. Vernon Riley, a cancer researcher in Seattle, about the effects of surroundings on the incidence of cancer among a particular strain of laboratory mice. The female mice harbored a virus that made them highly prone, under normal laboratory conditions, to develop breast cancers.

The usual conditions, according to the investigator, created chronic stress for the animals. The area was noisy, animals were frequently being handled by experimenters and blood was often drawn from the mice in the presence of the other mice.

When the mice were divided into groups, some in the usual environments and some in protected environments—quiet, less handling and no bleeding procedures performed in the room—Dr. Riley said the incidence of tumors dropped dramatically.

Further support, especially for the type of experiment conducted at Ohio State, came from a study reported in November 1982 in the *American Journal of Veterinary Research* by investigators at VPI. The report was prepared by Dr. W.B. Gross and Dr. Paul B. Siegel.

Their study involved three groups of chickens (a total of 1,370 were used): "socialized birds" that were touched and gently talked to or sung to daily by a person, beginning shortly after they were hatched; "ignored birds," that had minimal human contact, and "hassled birds" that at first were ignored, then were shouted at and otherwise disturbed by cage-banging and other loud noises.

The "socialized birds" had better feed efficiency, greater resistance to infection by a specific organism and a better defense reaction against a particular substance injected into them, compared with the "ignored birds," according to the researchers.

The "socialized birds" also had much better feed efficiency and antibody responses than did the "hassled birds." The "hassled birds," however, were less susceptible to infections by an injected agent than the "socialized birds," perhaps reflecting a type of adaptation to an adverse environment.

Exactly how positive environments promote positive bodily reactions is not entirely clear to the researchers.

But there are several implications of the experiments.

They may help explain why different researchers get different results on the same experiments when animals are involved and why some farmers have more success with their livestock than others.

But the overriding implication seems to be that kindness has biochemical consequences.

Mozart and the Starling

FEBRUARY 18, 1991

It's the case of Mozart and the starling: How did the pet shop bird know how to whistle a 17-note passage from a work that the composer had just completed a few weeks earlier?

Mozart himself may have inadvertently leaked the music, suggested an Indiana University psychologist who's an authority on the remarkable ability of starlings to mimic sounds that they hear.

Mozart, who habitually hummed and whistled, may have visited the pet shop a few weeks earlier and gave the notes away, long before the concerto was played publicly, Dr. Meredith West suggested at a meeting of the American Association for the Advancement of Science in Washington.

Starlings, after all, are great imitators. They can imitate cats meowing, roosters crowing, babies crying, water running, horns honking, doors squeaking and even hammers hammering, according to Dr. West.

They also can imitate human sounds, including words and whistled versions of songs, she told a session at the annual meeting of the AAAS.

Dr. West, along with Dr. Andrew King at Indiana University, have been studying for 10 years the mimicking abilities of the European starling, a ubiquitous black, speckled bird that many people regard as a nuisance because of its tendencies to flock and roost in huge numbers.

Starlings were introduced into the United States on an artistic note, Dr. West said. In the 1890s, a wealthy New Yorker decided to stock Central Park with all of the birds listed in Shakespeare's "A Midsummer Night's Dream.." He paid for the importation of 200 European starlings.

Now, Dr. West said, there are about 200 million in this country— about one per U.S. citizen. But Americans, she said, don't tend to appreciate what Europeans have long known about starlings, namely, their abilities to imitate various sounds.

In Mozart's time, in the 18th century, starlings were among trained musical birds that people kept as pets.

Not only do starlings mimic sounds, but they also can string together various sounds into what she described as long "soliloquies" that may include human speech interwoven among other sounds.

The birds also can recreate strings of connected events. They may, for example, imitate an alarm clock ringing, followed by imitations of clinking dishes and of people talking. Or, they may mimic the barking of a dog, followed by doors opening and closing, followed by a voice greeting the starling.

Dr. West said starlings seem to give back, in a sonar-like fashion, sounds from their environments, perhaps as a means of testing or probing reactions of people or other creatures around them. And, she added, their utterances are sometimes coincidentally comical.

She cited one case in which a starling exclaims, "I have a question!" as its claws are being clipped. Another became tangled in a venetian blind cord and started shrieking, "Basic research!" over and over.

Yet another would sniffle and say, "Hi." That was the mannerism of a person, who had an allergy, who took care of the bird.

Dr. West said that all of the starlings she and Dr. King have studied show interest in music or whistling.

She said they often produce rambling whistled tunes made up of songs that had previously been whistled or sung to them, "intermingled with whistles of unknown origin and other sounds."

One starling, she noted, would start whistling "Rock-a-bye baby," then switch to the rousing "William Tell" overture, and go back and forth between the two.

It's from that background that Drs. West and King enter into the case of Mozart's starling.

Mozart kept a diary of the money he spent, and on May 27, 1784, he noted the purchase of a pet starling. The entry also includes his notation of 17 notes of a musical score that the bird whistled, with the exclamation, "That was wonderful!"

Those 17 notes appear in the final movement of his Piano Concerto in G Major, according to Dr. West.

The concerto, she said, had been completed about six weeks before he bought the starling, and the work had not yet been played in public.

In fact, Dr. West said Mozart had sent a copy of the work to his father just the day before he bought the bird as insurance against someone stealing it and claiming it for their own. In those days, when there were no copyright laws, Dr. West said others would crib Mozart's work.

Then, a day after he sends the concerto to his father, he walks into a pet store and hears a starling whistle the 17-note passage from his new concerto's final movement.

"It must have been striking," Dr. West said. "So he did what any new creator would do—he bought the bird."

She said she and Dr. King suppose that Mozart had visited that, and perhaps other pet shops weeks earlier, and being the inveterate whistler that he was, he let the passage slip from his lips during the visit.

"Perhaps when you next encounter an assembly of starlings," Dr. West said, "you will, as did Mozart, stop and listen.

"And perhaps in their sounds, you too might hear salutations from one upstart species to another."

'For Nonconformity the World Whips You'
JULY 28, 1974

Lest anyone thinks that no fascinating, scientifically unfathomed mysteries yet remain in the commonplace, consider the phenomenon of imitation among living things, from insects to humans.

Cows grazing in a pasture generally tend to be aligned in the same direction.

Ducks that are suddenly flushed will regroup in an instant and fly off in the same direction.

Some species of fireflies, in certain parts of the world, will flash in unison.

Schools of hundreds or even thousands of fish travel as a unit, sometimes with the entire unit changing direction in an instant.

And, of course, people imitate. There is the phenomenon of fashions, where hair styles, skirt lengths and other kinds of "looks" often seem to pass like a quickly moving wave across the country.

The same may be said of fads and crazes that do not seem to be unique to our times.

"Men, it has well been said, think in herds; it will be seen that they go mad in herds, while they recover their senses slowly, and one by one," said the 19th century Scottish writer, Charles Mackay.

People also become ill in herds, not only during epidemics but also from seeing other people become ill. One notable example occurred at an Alabama school last year when a wave of illness swept through the school, class by class, within a few hours, finally affecting 105 students and teachers.

A public health investigation of that incident ruled out possible physical causes of the outbreak, such as food or water contamination, and concluded it was a case of mass hysteria.

Interestingly, the phenomenon of imitation in nature has captured the attention of two Washington, D.C. physicists, Earl Callen, professor of physics at American University, and Don Shapero, assistant professor of physics at Catholic University.

"Usually, we find a biological reason for imitation," they say in an article in the current issue of *Physics Today*, a publication of the American Institute of Physics. "There is safety in numbers, and when N [a given number of] animals swarm together the chance of an animal's being eaten by a predator is, if the animal looks and behaves like the others, only one in N, but considerably more if it 'stands out in a crowd.'"

In the case of fireflies, they noted, it is the males who send out the flash to attract females. If all the males flash simultaneously the light will be seen over greater distances and be more easily recognizable, by the females.

But descriptions are not explanations, especially for exactly how imitative behavior is communicated among the individual members of a group that behaves as a single unit. How, for example, does an individual gnat know how and where the cloud is going in the next instant?

Callen and Shapero do not suggest a precise mechanism, but they do draw a mathematical analogy between imitative behavior and the phenomenon of magnetism. "We suppose that most physicists have, at one time or another, while watching fish or birds or people, remarked on the evident analogy between social imitation and magnetism," they said.

Magnetism arises from the basic particles of which atoms and thus matter are composed. When those tiny particles, with their individual magnetic properties, are similarly aligned in a given substance—aligned like grazing cattle or the members of a school of fish—the substance will have certain overall magnetic properties.

But there are things that can affect the alignments, or the imitative behavior, of the basic particles of matter. Temperature, for example, can cause the alignments to flip or otherwise change, thus altering a given substance's magnetic properties.

Callen and Shapero delve into some heady equations and concepts to develop their analogy between magnetism and the imitative behavior of various living creatures. The point of their effort is to develop a mathematical-physical way of exploring what happens when individual "single particle" behavior coalesces into group, collective behavior.

One attractive thing about their efforts is the implication of a deep unity in nature—that there may be something analogous between magnets and fish and men.

On the other hand, there may also be the implication that there is a natural tendency for conformity (which humans sometimes like to believe they can rise above). If nature has built in such a tendency, then the message would seem to be: Alignment with the group takes little effort, while going out of alignment takes some doing.

"For non-conformity the world whips you with its displeasure," observed Ralph Waldo Emerson many years ago, long before men thought of equations that are figuratively saying something similar.

A Sky Full of Explanation Points
AUGUST 16, 1992

A few months ago, in a marina near Key West, I spent some idle moments watching large schools of small silver-sided fish dart nervously about, in group unison, between a pier and a tethered sailboat.

What attracted me to them was their concerted behavior; the massive schools acted as single organisms as they glided through the water, turning, diving and rising as if the individuals were all rigidly connected to one another.

Sometimes, something would cause a minor explosion in the midst of a school, producing momentary chaos as individual fish scattered a short distance from their original positions in the unified whole. But within seconds, they regrouped and the school became a single moving organism again.

That's not a unique phenomenon. We see similar things happen to flocks of birds and swarms of gnats. They fly as a unified mass; something scatters them, and within seconds, they come back together again as a unified mass.

But the Key West silver-sided fish do even more, according to a friend there who's knowledgable about the flora, fauna and food chains in the lower Keys.

When a barracuda is in the area, he said those little fish organize around the predator, at some distance from it and in a shape that's dictated by risk. The "hole" between the fish and the barracuda is widest at the predator's mouth end and narrowest at the tail end, where the threat of being eaten is the least.

How do those fish, fodder for bigger creatures, know how to do that? How do they "decide" who goes where in the formation. How is the safe distance from the barracuda's mouth determined? How do the individuals know their place in the larger pattern of the school in that case, and when they regroup following a momentary scattering?

Such behaviors reflect some deeply basic organizing principles in nature that scientists, for all their accumulating knowledge, continue to have a hard time explaining.

It's probably safe to say, in fact, that no one really knows what the organizing principles are for the Key West silver-sided fish schools, or for any other coherent group of living entities for that matter.

The most awesome example is ourselves.

Our bodies are made up of trillions of living cells, organized into specific colonies that carry out specific duties. There are liver cells, heart cells, kidney cells, blood cells, skin cells, bone cells, brain cells and many others.

A major mystery is how all of those types of cells arise in the first place.

We each start off life as a single fertilized egg, hardly larger than the period at the end of this sentence.

That cell divides into two cells; each of the two divide to form four cells, and so on. About a week after it all starts, a hollow sphere of cells becomes seated in the mother's uterus, and the special colonies of cells begin arising.

At some point, one cell becomes the ancestor of the cells that become the liver. Another becomes the progenitor of the heart's cells. Yet another becomes the parent of bone cells.

And nobody yet knows all of the details of how such organization comes about.

Dr. Lewis Thomas, the physician-essayist, says we should all be absolutely amazed at the process that occurs in the development of every baby.

The mere existence of a single cell whose progeny becomes the human brain, for example, "should be one of the great astonishments of the earth," Dr. Thomas wrote in an essay, "Embryology," that's included in his book, *The Medusa and the Snail*.

He said that people ought to be in a continual state of wonderment over such parent cells and over such currently inexplicable processes of organization.

Dr. Thomas said that if anyone should figure it out in his lifetime,

"I will charter a sky writing airplane, maybe a whole fleet of them, and send them aloft to write one great exclamation point after another, around the whole sky, until all my money runs out."

And since that whole mysterious organizational process came together to produce our first grandson 10 days ago, I'm willing to help pay for some of those exclamation points in the sky.

Fly Eye Study Provides Look At Gene Control
APRIL 6, 1995

Some scientists in Switzerland recently told of astonishing experiments they conducted on fruit flies—they caused the flies to develop eyes on their legs, on their wings and on their antennae.

The scientists did so by turning on the gene that causes eyes to be produced—a gene that resides within every cell in the flies' bodies but which normally is turned off in most of those cells.

Although their research might sound like grisly grist for sensational newspapers , they actually are studying something fundamental about genes and how they are controlled.

They are studying something so fundamental about life—the things that cause genes to turn on or turn off—that it touches on about every activity in our bodies. Our physical well-being depends on how our genes are controlled. Many of our illnesses, such as cancer, depend on such controls, too.

For just about all of human history, people could only regard living things much as they viewed the face of a mechanical clock, without having the slightest notion about the gears, pawls and springs behind the face.

In the past four decades, scientists have been discovering the equivalent of the works behind life's face, and the equivalent of the gears, pawls and springs turned out to be DNA, RNA and the myriads of proteins that are produced under the direction of those master molecules.

Genes, the storehouses of instructions that reside in nearly every cell of our bodies, are made of DNA molecules. The instructions are written in a relatively simple chemical code in the double-spiraled DNA molecules.

Scientists have learned how to read the code, and in increasing numbers of cases, they are learning how particular genes are linked to particular functions. They also are learning how specific genes are tied to specific diseases.

So much has been learned in recent years, in fact, that a leading cancer scientist was able to note, almost offhandedly, at a recent American Cancer Society seminar in New Orleans:

"We now know what causes cancer—at the cell stage." Dr. Edward Sondik, acting director of the National Cancer Institute, quickly added, "But we don't know why."

All cancer, Sondik said, is genetic. That doesn't mean that all cancer is passed along through genes. But it does mean that specific genes, either ones that are passed along or ones that are altered by radiation, chemicals or other agents, seem to be involved in the development of cancer.

Again, as in the case of the fruit fly experiments of the Swiss scientists, the key issue seems to involve genes that are switched on or off. Every cell in our body contains the genes to make every structure within us. A liver cell contains the genes that could direct the production of a kidney, for example, but that kidney-making gene is turned off in liver cells—just as the eye-making gene in fruit flies is turned off in the cells in the flies' wings, legs and antennae.

According to recent research, many cancers result when certain genes are lost or inactivated. Those are genes that seem to be responsible for suppressing tumors. In other cases, certain genes may be turned on to promote the wild proliferation of cells.

Genes and the things that control them are the most basic secrets of life, accounting for the amazing diversity of organs, tissues—and creatures.

"Don't forget," said Dr. John Laszlo, the Cancer Society's vice president for research, at the New Orleans meeting, "the caterpillar has the same genes as the beautiful butterfly. The only difference is which genes are turned on."

Vultures' Graceful Soaring Belies Their Lifestyle
March 29, 1987

Some things are better appreciated at a distance, like vultures.

In the sky, they are the essence of grace as they soar with seeming ease in lazy, peaceful circles. It's on the ground, close up, that the image of grace and beauty is destroyed. For vultures really are rather unattractive physically, and so are elements of their lifestyles, such as their diets. They eat carrion, a polite term for decaying flesh of dead creatures.

One local authority said there are reports that vultures will occasionally kill a small creature to eat, but that behavior is not the norm. Mostly, their food is acquired by more passive means, namely, circling peacefully in the air until they spot it or pick up its scent.

What brings vultures to mind is the annual news item about them returning to Hinkley, Ohio, around mid-March. Actually, according to Dr. Charles R. Blem of Virginia Commonwealth University's biology department, there's nothing particularly special or remarkable about the Hinkley event; a number of migrating birds return to the same area around the same time each year.

Besides, the vultures of Hinkley really don't migrate that far anyway. It's more like a shift northward of a few hundred miles, according to Dr. Blem.

Also, he noted that vultures are often incorrectly called "buzzards." Buzzard, he said, is a general term that particularly refers to members of the hawk family. In Europe, in fact, Dr. Blem said that when people talk about buzzards, they are referring to hawks.

The vultures that return to Hinkley are turkey vultures, which represent one of the two types of vultures that are prevalent throughout the eastern part of the United States. The other is the black vulture.

The turkey vulture has a naked red head; the black vulture's head is black. The turkey vulture has a long tail; the black vulture has a stubby tail. When soaring, the turkey vulture's wings form a shallow V; the black vulture's wings are flat when it is soaring.

The soaring of vultures is perhaps their most intriguing feature. As they float along, without flapping their wings, they normally sink slowly unless they are in air that's rising faster than they are sinking.

There are several sources of rising air that vultures may use. One is rising bubbles or columns of air that are caused by the sun-warmed ground. Another results from the wind blowing across hills, ridges and slopes. And another is a wave pattern set up in the atmosphere by wind blowing across obstacles like mountains.

Of those, the major sources of lift for vultures are the rising bubbles of air, called thermals, that often give rise to fluffy, cauliflower-shaped cumulus clouds. Making use of rising air is obviously an energy-saving move for vultures, especially the large and heavy vultures of East Africa. They don't have enough muscle power to fly on their own, so they have to have to use thermals or other sources of lift to keep aloft, to look for carcasses.

Whether it's by sight or scent that vultures detect their food from aloft isn't clear. It is known that turkey vultures have a rather keen sense of smell, according to Austin L. Rand, in his book, *Ornithology: An Introduction*.

He noted that gas line maintenance workers in Southern California have long known of the olfactory abilities of turkey vultures and routinely make use of that knowledge. A smelly substance is added to gas so that homeowners can quickly detect a leak and turkey vultures are strongly attracted to that odor. So, when a leak in a gas line is reported, the maintenance men drive along it until they spot a group of vultures gathered around the pipe. Generally, that's where the leak is.

Tons of dead and decaying animals are taken care of by vultures, and according to a study several years ago by John S. Coleman, then a fisheries and wildlife sciences graduate student at Virginia Tech, farm carrion is an important food source. That study was conducted in the area of the Gettysburg National Military Park, where, according to local legend, turkey vultures were first attracted after the Battle of Gettysburg because of the thousands of dead horses.

Rand, in his book, said that vultures can be sustained a day or so by what they can consume from a carcass in a relatively short time. He said some vultures have been observed to perch 24 straight hours or so without leaving for food. They also may leave the perch and soar; soaring, Rand suggested, may be an alternative for perching.

"But who can tell whether a soaring vulture is simply passing the time until it is hungry again or is watching the ground for a meal to be used immediately or to be remembered and used later?" Rand said.

Whatever the reason, they and other soaring birds they have a certain kinship with us. As noted by C. J. Pennycuick in a 1973 *Scientific American* report on soaring flight in vultures:

"Like man, soaring birds extend their powers of locomotion by using a source of energy external to their own bodies; they are perhaps the only other group of animals that do so."

Points of Light Help Fireflies
In the Quest for Preservation
AUGUST 2, 1992

It's one of nature's showiest semaphore systems, where flashes of light in the darkness do the communicating. Flash-flash-pause, a signal might go. Flash-pause-flash might be the answer.

The creatures doing the signaling are fireflies, also known as lightning bugs. Behind the exchanges, as is usually the case in nature, is the instinct to preserve the species. Evening talk is about reproduction.

Scientists say fireflies blinking in the back yard after the sun goes down are mostly male fireflies sending out signals for females. Males of a particular species speak with a particular pattern of flashes. The female responds with a specific pattern, usually less complex than the male's and one that only males of that species recognize.

Biologist Sara M. Lewis of Tufts University notes, in fact, that if you watch fireflies and figure out the female's pattern of flashes, you can cause males of that species to come to you, even alight on you, by flashing a pen light according to that enticing pattern.

It isn't easy being a firefly Casanova.

Some years ago, Dr. James E. Lloyd at the University of Florida detailed just how difficult it is, based on studies he conducted in some Florida meadows.

On any given evening, he said, males on the prowl outnumber females by ratios as high as 50-to-1. So competition is high, and a male firefly may search a long time before finding a mate. Also, according to Dr. Lloyd, fireflies are sexually active only about 15 to 20 minutes each evening, starting about 20 minutes after sunset.

In a study he reported in *Scientific American* in July 1981, Dr. Lloyd tracked 199 male fireflies, only a few of which found females of their own species and mated with them.

The average male firefly, he concluded from his study, travels between one-half to two-thirds of a mile and emits 455 flashes during a typical evening's search. And, Dr. Lloyd calculated, the chances are about 1 in 7 that a male will find a mate on a typical evening.

Or, put another way, the average male firefly makes seven evening outings before mating.

But there are false alarms.

Some females of other firefly species will answer a male's signals with flash patterns that mimic those of his own species' females.

In at least one instance, the motive is dinner—and nothing else.

Those *femmes fatales* of the firefly world, noted Dr. Lewis recently in a telephone interview, take positions on or near the ground in an area where males are flying.

They mimic the flash patterns of females that the males are seeking. When a male responds, the mimicking female devours him.

That's not the only hazard that male fireflies face, added Dr. Lewis.

Because males tend to fly higher and farther than females, she said, they are more vulnerable to other dangers, such as being eaten by bats or getting caught in spider webs.

Above and beyond nature's toll on fireflies, Dr. Lewis is concerned that something else is threatening the insects. There are indications, she said, that fireflies may be on the decline in the United States.

If so, we may be to blame.

Dr. Lewis said she suspects the reason for the decline is that habitats for fireflies are being destroyed in the name of development. Formerly dark rural areas are being converted into housing developments and shopping centers. Woods are being cleared. There are fewer open fields, which are prime breeding grounds for fireflies.

And there are more and more street lights.

Fireflies, by virtue of their light-based communication system, need to talk in the dark.

COSMIC

There is no unknown. There are only things
not yet revealed, not yet understood.
Capt. James T. Kirk, *Star Trek*

One of the most astounding developments in science during the past four decades has been support for a grand idea about how our universe came into being. By the early 1960s, astronomers were divided over two major theories of the origin of the universe, the "steady state" theory and the "big bang" theory. The "steady state" theory suggested that the universe always existed and always will, while the "big bang" theory suggested that the universe had a definite beginning, in the explosion of a primordial ball of energy some 10 to 20 billion years in the past. Each theory accounted for a central observation about the universe, namely, that it is expanding; galaxies are flying away from one another.

A major proponent of the "steady state" theory was the noted British astronomer, Fred Hoyle, who told me in an interview at Virginia Tech in December 1964: "I think that the 'big bang' thing is not taken seriously anymore." Less than six months later, two scientists at Bell Laboratories, Arno Penzias and Robert Wilson, said they had detected a uniform, weak static in the universe that fit predictions derived from the "big bang" theory. They had discovered, through a radio telescope originally built to handle signals from the earliest communications satellites, the diminished energy from the birth of the universe. The "steady state" theory, for all of its merits, faded as the "big bang" theory, bolstered by a vital piece of real evidence, began to flourish. At the same time, scientists were seriously considering the ultimate fate of the universe—will it go on expanding forever, or will it eventually collapse and trigger another "big bang," and pulsate from "big crunch" to "big bang" ad infinitum?

Scientists continue to consider such questions. They have vastly refined the "big bang" theory, which is still a work in progress, and they are accumulating additional insights into the early universe through the

Hubble Space Telescope and other sophisticated instruments.

The following columns reflect something of the evolution of cosmic thought from the mid-1960s to more or less the present. They also include a hint—through the report on an address by astronomer Virginia Trimble to the Richmond Physics Club in the 1970s—of what is now sometimes called the anthropic cosmological principle, namely that the emergence of intelligent beings like ourselves seems to be a consequence of the universe's fundamental laws and forces.

Readjustment May Soon Be Necessary
JUNE 20, 1965

As if there's something in the air, a number of developments that bear on the nature of the universe have been streaming from observatories and laboratories in recent weeks.

From all indications, these recent events could be forerunners of a unified view of the universe as far reaching, if not more so, as the discovery in the 1920s that the universe is expanding.

In fact, if the recently announced preliminary reports are ultimately borne out, there will undoubtedly be a major readjustment of thinking about our place in the universe.

To recapitulate, the recently announced developments include:

(1) Bell Telephone Laboratories' scientists reported that they have apparently detected—in the form of radio waves—the flash of energy that occurred when the universe was born.

Time and space have long since stretched that energy into short radio waves, according to a theory that was being developed at Princeton about the same time the Bell scientists picked up their strange hum of radio noise.

For a short period, the Princeton group was not aware of the Bell group's discovery, and the Bell scientists—who detected the unexplainable radio waves on a large horn antenna normally used for communications satellite studies—were unaware of the Princeton theory. The two groups getting together resulted in the announcement about a month ago that the original birth flash of the universe had seemingly been detected, thus giving new weight to the theory that the universe began with the explosion of a primordial fireball.

(2) Preliminary studies of two kinds of bright, recently discovered

objects in the skies indicate, according to one astronomer, that not only did the universe begin with a "big bang," but also that it is finite and pulsating, almost like an extremely slow heartbeat.

About a month ago, Dr. Allan Sandage of Mount Palomar Observatory reported that his study of nine quasi-stellar sources—extremely distant, bright objects—indicates that the universe is a closed system that pulsates once every 82 billion years.

Last week, Dr. Sandage announced the discovery of a "major new constituent of the universe"—blue, extremely bright galaxies that resemble quasi-stellar sources (or quasars) in many respects.

Both quasars and the newly discovered blue galaxies are so far away that looking at them is equivalent to looking billions of years into the past. Thus, by studying these objects, one ought to get some idea of how the universe has changed over a major part of its lifetime.

This, in turn, ought to yield clues about how the universe is behaving, and whether it is expanding into infinity, or whether its surface is closed like that of a sphere's.

Dr. Sandage's preliminary evidence, based on his studies so far of quasars and the blue galaxies, suggests that the universe is closed and alternately expands and contracts once every 82 billion years, he states.

That is, the universe expands, slows down, then begins contracting. It contracts to become a fireball-like object again, and again, a gigantic explosion begins anew the expansion process. The universe pulsates like this indefinitely, according to the theory.

(3) The nucleus of an anti-atom has been produced in the laboratory, a team of Columbia University scientists reported last week, thus adding weight to past speculations that anti-worlds may exist in the universe.

In anti-worlds, atoms have negative nuclei with positively charged electrons spinning around them, instead of positively charged nuclei and negative electrons as it is in our world.

Although the existence of the anti-particles—particles with the reverse characteristics of the particles that make up our world—has been known for some years, it has not been known for sure whether anti-atoms could be built from them.

The finding announced last week, according to the study's authors, demonstrates this. Further, they say, it means no one can rule out theories of anti-stars, anti-planets and even thinking creatures composed anti-matter.

For the present, the pulsating universe theory—and the reported evidence for it—may have the most profound effect on our thinking about our place in the universe.

If subsequent data back up this theory, it will mean for one thing that our universe is not infinite; we live in a universe with boundaries. We have a mind capable of realizing infinite space, but the grandest physical structure we can comprehend falls short.

Further, we would have to accept the fact that every 82 billion years, or whatever the figure might turn out to be, the slate is wiped clean—everything is destroyed in the melt of a contracted, gas-like fireball.

If the universe is expanding into infinity, and would always do so, there may be localized deaths but there would be overall existence.

With a pulsating universe, however, we—and all intelligent forms elsewhere in the universe—would have to live with the knowledge that all is destined to vanish.

Universe's Noise Discovery Backs Creation Theory
DECEMBER 31, 1978

It was the year of the test tube baby, paraquat and a sensational book about the alleged cloning of a human being.

It was a year in which the "carcinogen of the week" syndrome continued to persist; when the word came that two aspirins a day might help prevent stroke in some people, and when probes were sent to Venus that raised additional questions about that member of the solar system.

And it was a year in which a fundamental discovery about the nature of the universe was recognized through part of the Nobel Prize for physics. And that, perhaps, is worth a bit of reflection as the year ends.

Part of the Nobel Prize in physics this year went to two Americans who, in 1965, discovered compelling evidence that our present universe had a beginning, a birth date, a moment when its age was zero.

The discovery of Arno A. Penzias and Robert W. Wilson of Bell Laboratories was basically this: They detected a background noise—a kind of uniform static—coming from the sky that seemed to be emanating equally from all directions.

Their discovery was accidental. They were using an antenna built some years earlier at Holmdel, N.J. for sending and receiving signals to and from the pioneer communications satellites Echo and Telstar. Penzias and Wilson were using that antenna to conduct certain radio astronomy studies of our galaxy, the Milky Way, when they found the little bit of noise that they could not account for.

Meanwhile, a group at Princeton was figuring there should be such a noise coming from the universe and, in fact, efforts were beginning there to search for it. Further, some theoretical physicists back in the 1940s suggested that such a noise should be present if the universe had a particular, unique beginning.

All of those thoughts were based on the theory that the universe began with a tremendous outburst of a ball of dense, hot material. That theory, incidentally, is popularly called the "big bang" theory of the universe's origin.

And what Penzias and Wilson inadvertently detected was the leftover "noise" of the explosion in which the universe—the galaxies, the stars, planets around the stars and, indeed, ourselves—was created. That was the interpretation of the Penzias-Wilson finding at the time, and subsequent studies have strengthened that interpretation.

Their discovery provided a major bit of evidence that helped settle a scientific and philosophical controversy that had existed for nearly 20 years. The controversy was over two major ideas about the nature of our universe as far as time, or age, was concerned.

One idea was embodied in the "big bang" theory, namely, that there was a unique moment of creation for the universe. The other was that the universe had always existed and will continue to exist in its present form. Individual stars and galaxies may come and go, just as individual particles of water come and go over a dam, but the universe as a whole will continue, just as the stream behind the dam remains more or less constant. That theory of an infinite-time universe, with no unique moment of creation, was known as the "steady state" theory.

Around the time Penzias and Wilson were trying to track down the cause of the puzzling omnidirectional radiation, theoretical astronomers and physicists appeared to be equally divided over the two theories. Some indicated they personally and philosophically preferred the idea of a continuous, "steady state" universe; others leaned to the idea of a "big bang" beginning.

The mid-60s discovery by Penzias and Wilson was compelling evidence in favor of the "big bang" theory and a serious blow to the "steady state" theory that was being advocated at the time.

So, today, if you ask astronomers how the universe began, most will reply that it started at a given time with the explosion of a hot, dense ball of material. And they are likely to mention the discovery of Penzias and Wilson as the chief reason for that belief.

At least, their discovery was a turning point in the history of modern ideas about the nature of the universe.

As the Swedish Academy noted upon choosing Penzias and Wilson for the 1978 Nobel Prize in physics (one-half of the prize was given to a Soviet physicist for his low-temperature studies):

"The discovery of Penzias and Wilson was a fundamental one: It has made it possible to obtain information about cosmic processes that took place a very long time ago, at the time of the creation of the universe."

It will take the traditional test of time to determine how significant other scientific events during 1978 will turn out to be, but at least it was a year in which two scientists were recognized for learning something basic about our existence.

And that is that we live in a universe that had a unique, definite and violent beginning.

No one yet knows for sure whether the whole thing will collapse someday into a tight, hot ball to explode again, thus starting another universe, or whether our present universe began that way.

Nevertheless, the Penzias-Wilson discovery has helped clear the view for scientists trying to answer an ultimate question.

A Sunday Morning 'Service'
FEBRUARY 21, 1993

It was 8:30 on a Sunday morning, at a science meeting in Boston. The room for this session was filled; all 150 or so seats were taken, and some people were standing.

The topic that brought them out so early on a Sunday morning was seemingly an unusual one for a science meeting: "The Religious Significance of Big Bang Cosmology."

Behind the session, one among many on various topics covered at the annual five-day meeting of the American Association for the Advancement of Science, were some recent findings related to the beginning of the universe some 12 to 15 billion years ago.

The findings, from an orbiting observatory called the Cosmic

Background Explorer or COBE, have been described as the closest in time yet to the actual creation of the universe, which occurred in an explosive instant that astronomers now call simply the big bang.

Microwave detectors on COBE searched for, found and painted for their designers some abstract, data-rich pictures based on the radiation the universe gave off when it was only 300,000 years old. When you look up at the stars on a clear night, you are seeing light that left them years ago and just now is arriving here. Looking up is looking back in time, seeing things not as they are this instant but as they were when the light you see left them.

If you have the right kind of instruments at the right places, you ought to be able to see well back into the beginning of time, or close to it.

And that's what COBE did and still is doing. The picture it yielded of the universe when it was about 300,000 years old is roughly equivalent to looking back in a 70-year-old person's life to when he or she was less than a full day old.

Until COBE's recent data, the closest astronomers could get to the creation event was several billion years, through observations of distant quasars. That's like looking back to the youthful years of a 70-year-old.

In a widely quoted remark, criticized by some but nevertheless remembered by many, Dr. George Smoot, the leader of the COBE microwave experiment team, said of the splotchy picture the ancient radiation painted: "If you're religious, it's like looking at the face of God."

Dr. Joel R. Primack, a theoretical physicist at the University of California at Santa Cruz and one of several speakers at the Sunday morning session, suggested there may, indeed, be some religious significance behind COBE's data—especially for theoretical scientists whose work is touched by the findings.

He described what has been known about the universe's beginnings, and how scientists know what they do know about it. Astronomers can see, by their photographic plates of the galaxies, that the universe is expanding, as if in the midst of a giant explosion. They can detect the radiation the explosion gave off, now a uniform hiss of feeble static on radio astronomers' detectors.

And the abundance of some of the lighter chemical elements in the universe fits predictions that are based on the premise that the universe began with a big bang.

There are more recent theories, Dr. Primack explained, that during the first few trillionths of a second into the creation, the universe—then billions of times smaller than a single electron in our bodies—underwent an accelerated expansion.

At the end of that accelerated expansion, when the universe was about 5 inches across, the universe settled down to the regular expansion that has been going on ever since. The ripples detected by COBE may be related to that early inflationary period of the universe's birth and formed the seeds of the galaxies of stars that came later.

Now there are certain features of the present universe that suggest most of the matter in it still has not been detected directly by scientists. They still don't know exactly what it is. Dr. Primack is one who has proposed a theory about the universe's dark matter and made some predictions about its density.

COBE's data, it turns out, are in line with Dr. Primack's predictions. And therein lies part of the personal religious experience that COBE helped provide for him. He said that he himself, and "most other theoretical physicists I know, think of themselves as religious."

"Our religion is peculiar," he added. "Our faith is that the universe is intelligible and at the bottom is beautifully simple. And we have religious experiences, as when we make predictions that turn out to be true."

When that happens, as occurred when COBE's data agreed with his predictions about cold dark matter, "It makes you think you are in touch, in a deep sense, with the universe."

If the Universe Were Different, So Might We
NOVEMBER 27, 1977

There's a curious thing about the universe in which we live. It almost appears that its design is such that life was likely, if not bound to appear.

Put another way, if any one of a number of the universe's major characteristics were different, chances are that we wouldn't be here—vitally aware of the whole thing and wondering about the existence of other beings who also are aware.

It's entirely possible, according to astronomers and physicists who have been pondering the nature of our universe, that our universe didn't have to be the way it is. It could have been different. And if it had been different, life would have been impossible—at least, the chemical-based life that we know.

From such ponderings, which reach into realms where distinctions between science and philosophy become blurred, emerge the possibilities that many other universes exist, or have existed, and none of them had any chemical-based life whatsoever.

The grand sweep of our universe and its curious nature was succinctly outlined by a bright, wry astrophysicist recently in a lecture before the Physics Club of Richmond. She is Virginia Trimble, who has been named outstanding young scientist by the Maryland Academy of Science and who holds teaching appointments at the Universities of Maryland and California.

Since the story of our universe, like all good stories, should begin at the beginning, Trimble traced the universe's history from time zero—about 20 billion years ago—"down roughly to the birth of Richard Nixon," as she put it.

The universe—the totality of all the galaxies of stars that astronomers can detect—presumably began with the explosion of a huge, dense ball of material. It is meaningless to ask what preceded the dense, hot ball, since all past characteristics would have been wiped out by the extreme temperatures in the ball.

Trimble's analogy was this: If you throw a Cadillac into a blast furnace, it would be impossible to tell, by analyzing the molten metal that flows out, whether it came from a Cadillac or a Rolls Royce.

So we can't tell what went on before our present universe began violently some 20 billion years ago from a tight ball of material and energy. But we do know that the universe, once the primordial ball began expanding, consisted of a lot of hydrogen and a little bit of helium early in its history.

And, fortunately for us, the material of the universe tended to be lumpy, rather than spread out evenly throughout space. If all the matter in the universe were spread evenly, "You would have been a lone hydrogen atom and you would never meet another one in your lifetime," Trimble said, noting the average density of the universe is one atom in about a quart of space.

As it turned out, however, the lumps of the early universe—lumps of gas and dust—became galaxies with billions of stars in them. And the stars, in turn, became the furnaces in which elements beyond hydrogen and helium were formed.

The iron in our blood streams, the carbon on which life is based, the oxygen we breathe, and the other elements essential to our existence were formed within stars. Some of them blew apart, spewing those elements into space, from which new stars and planets like our own were formed, and through which those elements eventually found their ways into cells and creatures.

Trimble recounted experiments over the last 25 years that have shown how easily the building blocks of life can be formed with a few primitive gases, composed of some of those spewed-out atoms of carbon, nitrogen, hydrogen and oxygen, and energy. Further, the processes that produced a brownish slime of such building blocks for life in the laboratory appear to be going on in space, she noted, referring to recent discoveries of rather complex organic molecules in interstellar dust clouds. "Apparently, chemistry says that where you have these compounds and energy, the process that leads from a slime mold to a politician is inevitable," Trimble quipped wryly.

Or, perhaps less prosaically, it seems that chemical-based life is a "very natural consequence" of the way our universe is arranged.

The fact that life is possible in our universe actually goes deeper than the chemical reactions that are presumed to have led to living things. As Trimble noted, if any of the four known basic forces of physics were a little bit different from what they actually are, there would not have been life.

If the gravitational force, for example, were somewhat stronger that it is, stars would have burned hotter and their lifetimes would have been too short for life to develop. If gravity had been too weak, the stars would have burned too coolly to cook up many of the elements necessary to life.

If the electromagnetic force, the force involved in chemical reactions, were too strong or too weak, atoms as we know them would not have been possible. A third force, the nuclear force, could not have been too different; otherwise, the synthesis of many of the elements would not have proceeded as it did.

And the fourth force, the so-called weak force, was involved in the burning processes inside stars. That, too, could not have been too different, or else things would not have turned out as they have.

Trimble also referred to several other characteristics of the universe, including its rate of expansion, its density at the beginning and others, that are just right for the kind of universe in which chemical-based life could emerge.

If any single one of those various characteristics had been different, the result would have been a universe devoid of life as we know it.

What it all means, scientifically, is not at all clear. Therefore, there are speculative musings. It could mean that there are many possible kinds of universes and ours is but one. It could mean that our universe will eventually die in a black hole and another, difference universe will emerge.

It could mean that many universes co-exist in five dimensional space and we are not aware of them.

It most probably means no one yet knows why our universe is the way it is.

THE WAY WE ARE

Speak roughly to your little boy,
And beat him when he sneezes;
He only does it to annoy,
Because he knows it teases.
LEWIS CARROLL, *Alice's Adventures in Wonderland*

The human body, its care and feeding and quirks, has never failed to fascinate me.

I recall as a youngster being intrigued by a magazine account of how a frontier physician studied human digestion through a hole in a French trapper's abdomen that was made by a gunshot wound. Years later, during a vacation trip to Mackinac Island in Michigan, I was quite thrilled to find that the whole story, from the accidental wounding of Alexis St. Martin to his treatment and study by Dr. William Beaumont, took place right there, at the very place I was visiting. When I returned from vacation, I wrote "Stomach 'Window' Made History."

My own body was the basis for one column. During a feverish bout with the flu, I intermittently wondered about what was going on within me that made be feel so miserable. Possible column material, I thought. I remembered skimming through a recent book about the immune system, and I resolved to call the author when—and if—I recovered for an explanation of what goes on when our bodies are invaded by flu viruses. "Body A Battlefield During Flu" was the result.

God Bless You: The Anatomy of a Sneeze
AUGUST 14, 1983

It may begin with an irritant, such as some of the weed pollens that become prevalent this time of year.

In the nose, they trigger nerve impulses that flow through the trigeminal nerve to brain structures at the base of the brain, in the region where the brain and spinal cord merge.

From there, signals flow back to a nerve center in the face, from which impulses radiate to glands and blood vessels in the mucous membrane of the nose. The membrane becomes engorged with a clear nasal secretion. The sensation, incidentally, is not entirely uncomfortable. In fact, it may be rather pleasurable.

The secretion into and the swelling of the nasal membrane, meanwhile, irritate the trigeminal nerve, and impulses again flow to the structures at the base of the brain, the pons and the medulla. The medulla contains, among other things, the respiratory center, plus various mechanisms for reflexes.

As a result of the respiratory center's stimulation, the inspiratory mechanism is touched off by signals carried by the nerve that controls the diaphragm. In plainer words, there is a deep inhalation.

Then, the palate raises, and a throat muscle contracts to shut off respiratory passages from the nose. Pressure builds up in the lungs, and is forcefully expelled through the throat and mouth. The eyes close involuntarily.

And someone is likely to say, "God bless you."

Therein lies the complex anatomy of a sneeze, a phenomenon that is as linked to the flowering of the late summer's ragweed as pupil constriction is to the bright outdoor sun.

As a matter of fact, the bright outdoor sun can also trigger sneezing in some people, through most of the same routes that are involved with the sneezes that are detonated by irritants in the nose. Strong odors can touch off sneezing in some.

So can fear. So can certain pleasurable stimuli. One medical journal article notes that "sexual excitement may provoke sneezing episodes." Sneezing has also been associated with seizures arising in the temporal lobe. And stimulation of the temporal lobe during surgery has been known to produce a sneeze.

Sneezing apparently occurs among most if not all mammals, according to Dr. Brent V. Stromberg, a Johns Hopkins Hospital surgeon who reviewed the matter of sneezing in a 1975 issue of *The Eye, Ear, Nose and Throat Monthly*. He describes sneezing as "a primitive protective function."

Aside from the complicated sequence of events involved in a

sneeze, there is also interest in how such a body phenomenon has become such a focus of myth and superstition. "Why did benevolent wishes like 'bless you' or 'gesundheit' attach themselves inseparably to the sneeze?" asked Dr. Selig J. Kavka of the Mount Sinai Hospital and Medical Center in Chicago, in a recent issue of the *Journal of the American Medical Association.*

Both he and Dr. Stromberg refer to some of the rich lore surrounding the sneeze. It is said, for example, that Prometheus, the titan of Greek myth who stole fire from the gods and gave it to humans, also introduced the sneeze to mortals. Most ancient cultures considered the sneeze to be of sacred significance, either ominous or favorable.

Adam is said to have sneezed for the first time when Eve tempted him with the apple, therefore suggesting that the sneeze is an evil omen. But when Jacob sneezed while asking God to answer his prayers, sneezing was considered to be favorable. Some cultures regarded a sneeze as an expulsion of an evil spirit.

"God bless you" is said to have arisen during the early part of the seventh century, during a plague in Italy. The story was that victims would sneeze a few times and die. As a result, the story continued, Pope Gregory ordered the people to utter the short benediction, "May God bless you," to sneezers.

Dr. Kavka suggested that much of the lore indicates a danger inherent in the sneeze, and indeed, modern medicine recognizes certain ones, including the spread of disease-causing organisms. Also, sneezing has occasionally been known to cause blood vessel damage, especially if the sneeze is forcefully smothered. Nosebleeds might be one result. "A rarer, but occasionally disastrous consequence is a stroke following forced smothering of sneezing attacks," wrote Dr. Stromberg.

Thus, it's best not to try to bottle up the pressure behind such a violent event (the droplets expelled from a sneezer's mouth can have speeds of nearly 70 miles an hour). There may be good reason, after all, for comments about health and blessings following a sneeze.

Stomach 'Window' Made History
JULY 30, 1989

Mackinac Island is a resort and tourist attraction a few miles off-shore in Lake Huron, near its watery connection with Lake Michigan. You have to take a ferry boat from the Michigan mainland to get there, but if you go in mid-summer, you will have plenty of company.

The streets of the little town, you find upon arriving, are crowded with vacationing pedestrians, bikers and passengers in horse-drawn vehicles.

But hordes of people on this island are nothing new.

Two hundred years ago, Indian and trappers' canoes, plus other kinds of water craft, also brought large numbers of people to the island, where John Jacob Astor established the headquarters for his fur trade and where a strategic fort was built.

In the early 1800s, this busy little island was one of the most remote and isolated frontiers of American civilization.

Yet despite the island's remoteness and primitive conditions, a medically significant event occurred there nearly 170 years ago that is recounted in medical textbooks and journals even today.

It was there that an Army physician learned something about how our stomachs work.

The physician was Dr. William Beaumont, who in 1820 was assigned to Fort Mackinac as the post surgeon. In those days, at that frontier outpost, violence and a variety of diseases provided plenty of challenges for a doctor, one historian later noted.

One day, two years after Dr. Beaumont arrived, a trapper named Alexis St. Martin was wounded in the abdomen by a shotgun that had accidentally gone off. The accident occurred in a retail store in the town below the fort, where the only physician for hundreds of miles around was located.

Someone ran up the hill to the fort to get Dr. Beaumont, who came immediately to the store. St. Martin, the physician found, had two broken ribs, a rupture of the lower part of the left lung and stomach lacerations. Dr. Beaumont did not think that St. Martin would live, but he began treating the wounded man.

He cut off a sharp end of one of the broken ribs; he cleaned the wounds and he did what he could to repair the damaged organs.

St. Martin lived, and Dr. Beaumont gave him intensive care during the ensuing days, weeks and months. The physician even took the trapper into his own quarters at Fort Mackinac in order to nurse him back to health.

The man eventually recovered, but one problem remained. The hole into St. Martin's stomach never closed completely.

At some point, a year or two after the accident, it occurred to Dr. Beaumont that the opening in St. Martin's abdomen was a natural window through which he might observe human digestion.

Up to then, scientists were beginning to figure out—through studies involving vultures, fish, frogs, snakes, cattle, horses, dogs and cats—that food was digested through chemical processes, through the action of gastric juices.

Previously, people had speculated whether the stomach worked like a grinding mill, or like a vat in which fermentation took place, or, in the words of one 18th century physician, like a "stew pot."

St. Martin's opening gave Dr. Beaumont an opportunity to study digestion in a living human being. In her delightful, highly readable *The Great American Stomach Book*, Maureen Mylander noted that Dr. Beaumont would lower pieces of food, tied to a silk string, into the hole in St. Martin's abdomen and withdraw them after varying lengths of time to study the effects of the gastric juices on them.

Mylander said that every day for a year or so, St. Martin "endured meat, potatoes, bread, fruit and vegetables until, unstrung at last, he ran away to Canada."

But Dr. Beaumont was able to persuade the young trapper to take part in further experiments intermittently over the next few years.

In 1833, 11 years after St. Martin's accident, Dr. Beaumont published his findings in a book titled *Experiments and Observations on the Gastric Juice and the Physiology of Digestion.*

Today, nearly every account of human digestion, whether it be in a professional or a lay-oriented publication, refers to Dr. Beaumont's studies of St. Martin.

And on Mackinac Island, a visitor today can see where that bit of frontier medical research took place—and be impressed as I was that such a significant, lasting contribution to medical science could emerge in such a modest, remote location.

The Michigan State Medical Society acquired the American Fur Company retail store in which St. Martin was wounded, established a

Beaumont museum in it, and donated the building to the Mackinac Island State Park Commission in the early 1950s. Also, one can visit Dr. Beaumont's quarters at Fort Mackinac where he and his wife cared for St. Martin.

There's another ending to the story, incidentally:

St. Martin, who was 18 at the time he was shot, subsequently married, had 17 children and outlived Dr. Beaumont by 27 years. He was 76 when he died in 1880; Dr. Beaumont died at age 67 in 1853.

Body A Battlefield During Flu
JANUARY 5, 1987

It began insidiously. Some viruses, apparently flu viruses, invaded somewhere in the victim's upper respiratory system's mucous membranes.

The tiny entities—it's questionable whether they can be called "creatures" because they are raw pieces of genetic material, with no life of their own—penetrated some cells and, like terrorist hijackers on an airliner, took total control. They injected themselves into the cells' genetic machinery, causing each of the cells to make hundreds of copies of the viruses.

Then, the virus-laden cells began bursting, releasing new viruses, each bent on invading another cell. The whole invasion rapidly escalated.

At this point, the victim was unaware that an invasion was under way. But his internal defense system had detected trouble and was responding to it.

Constantly circulating through the body, in continual contact with every organ and tissue, are types of white blood cells called phagocytes that are on the lookout for anything unusual, whether it be bacteria, foreign particles inhaled in the lungs or viruses.

When patrolling phagocyctes come across the rapidly multiplying viral invaders, they subdue the foes by engulfing and digesting them—literally, by eating them. But, the job quickly becomes too much for the first-line defenders alone as new cells burst, releasing new hordes of viruses.

Another class of white blood cells, T-lymphocytes called helper T cells, comes on the scene. They take note of the type of invader that's being battled, and send out chemical alarms for other kinds of troops. Those include types of T cells called killer cells that seek out and destroy cells that have already been infected by the viruses.

The second line of defense also includes a type of white blood cell

called the B-lymphocyte, or B cell, which produces materials that neu-
tralize the invading viruses.

So the battle was being furiously fought in a variety of ways; the
enemy was being eaten, overwhelmed outright and disarmed.

It was only when the battle was launched that the victim knew he
was getting sick. "The symptoms (of the flu) were associated with the
response of your immune system," Dr. Steven B. Mizel told the victim
later. "They were due to the body's response to the attack" by the viruses, he said.

Dr. Mizel, a widely known authority on the immune system, is the
author of a recent book, *The Human Immune System: The New Frontier
in Medicine* (Simon & Schuster Inc.), that explains the workings of the
intricate immune system in layman's terms. He is chairman of the micro-
biology and immunology department at the Bowman Grey School of
Medicine in Winston-Salem, N.C.

The victim dragged his body, with the raging battle going on inside
it, to work, in the false belief that he would feel better as the day wore on.
But the aches did not abate; the skin appeared to become increasingly
sensitive; the headache worsened, the appetite had totally vanished, and
there was a washed-out feeling that made the victim want only to take the
raging battle to bed.

And once in bed, the chills were unrelenting, even with two blan-
kets and a heavy cover over him. He had no interest whatsoever at that
point in what his temperature was (or in anything else) but a few hours
later, when his wife measured it, it was 101 degrees Fahrenheit.

It's only been within the past few years, Dr. Mizel explained, that
the source of the unpleasant symptoms of an illness has apparently been
found. Most of them involve a substance released by the first-line defenders, the phagocytes. That substance is called interleukin 1, which helps
promote increased production of T cells.

It also causes fever. A few years ago, Dr. Mizel said, researchers
found that interleukin 1 travels through the bloodstream to the brain,
where it signals the body's temperature-control center to turn up the temperature a bit.

Why? Because, researchers later learned, the immune system functions better when temperatures are higher than normal. Phagocytes seem
to destroy their prey better at elevated temperatures, and T and B cells
grow better at higher temperatures.

"The whole thing is very tightly controlled," Dr. Mizel told the victim. Most people suffering an infection like that caused by flu do not develop fevers much above 101 degrees, he said. He also said that it's best, in such cases, "to let the fever take its course," rather than take aspirin or any other agent to reduce it. The fever is there for a purpose, according to Dr. Mizel, and that's to create an ideal environment for the immune system's fighters. "The immune system evolved over a very long period of time—several million years—and if you overtreat it, you may cause problems," he said.

Interleukin 1 also seems to be responsible for the muscular aches and weakness that flu sufferers, and patients suffering other kinds of diseases, experience. Dr. Mizel said that interleukin 1 causes the breakdown of muscle tissue, presumably so that the raw materials of the breakdown can be used to make the energy and other vital ingredients of the defense system.

Further, he said some recent studies indicate that interleukin 1 is responsible for appetite loss in illness. But it is not yet clear, he indicated, what the value of appetite suppression during an illness is.

The chills the victim experienced during the first afternoon of his apparent bout with the flu, Dr. Mizel said, were due to changes in blood flow in the body as a result of its temperature increase. There is, he explained, "an ebb and flow of the immune response," which means that the fever and other elements of it occur in regular cycles of six or so hours.

About mid afternoon, the victim's wife called from her workplace to check on him. At first, she did not recognize his voice, largely because of his head congestion. That, explained Dr. Mizel, was due to particular cells, called mast cells, releasing histamines and other substances that, he said, "let the immune system work better."

By 7 a.m. the following morning, the victim's temperature was slightly under 100 degrees. By midday, it was essentially normal. The battle was won by the good guys, the phagnocytes, the T and B cells under the intricately orchestrated control of the victim's immune system. Even the end of the battle was under exquisite direction.

Yet another class of T cells, called suppressor T cells, entered the scene and released substances that slowed and finally stopped the attacking cells.

And the victim had a cup of chicken soup.

'I Got This New Stuff and It Costs a Million'
JUNE 7, 1993

Admittedly, the numbers are small but the fact it happened at all is interesting. And so are the implications.

Dr. David L. Steed, a surgeon at Presbyterian University Hospital in Pittsburgh, spoke at a recent medical conference here about a study on wound healing. The study involved a dozen patients who had wounds—they included such things as bed sores, diabetic ulcers, cuts and surgical incisions—that would not heal properly.

The patients, he said, had the problems for an average of 20 months. Some had the non-healing injuries for as little as three months while others had them for several years.

In the study, the patients were first treated with a placebo and then with an agent designed to promote healing.

Three of the patients who were treated with the placebo, Dr. Steed said, experienced a complete healing of wounds.

A placebo is a medical blank; it may look like the ointment or pill that's being studied but it has no chemical or biological action. It is a sugar pill, a sham.

But the placebo—the word comes from the Latin meaning to please—can persuade. Through links yet to be mapped within us, the belief that something can help often is translated to an actual improvement in an ailment or condition.

As Dr. Steed noted, you can treat patients' chronic skin ulcers for months with little progress, until you give them something new with an incantation that goes something like: "I got this new stuff and it costs a million dollars a tube."

Placebos indeed have their advantages. He clicked on a slide listing some, such as "Inexpensive," "Readily available" and "Free of infection."

But placebos are not benign. They, too, can have side effects.

Consider some drug ads in recent medical journals that compare side effects of the drugs with side effects that people reported when they were given placebos.

An ad for a blood pressure drug showed that some people who got the placebo during studies of the agent reported having such problems as diarrhea, headache, dizziness, nausea or chest pain.

Another ad, for an antidepressant, showed that certain percentages of those who got the placebo became nauseous, fatigued, suffered insomnia or agitation.

Yet another ad, for a cholesterol-lowering agent, showed that some who got the placebo developed gastrointestinal upsets, indigestion and stomach pains.

For much of medical history, the placebo was the main thing that healers had to offer their patients. And that's not to say that they had nothing to offer. Quite the contrary. The placebo effect, while not perfect, is real enough to make many ills better—and to sustain the reputations of generations of medicine men and even medical doctors.

A typical good medical study today is one in which patients are randomly assigned to two groups. Where possible, members of one group get the agent that's being tested and members of the other group get a placebo, an agent that's made up to look, smell and taste like the real thing.

In order to rule out the power of suggestion, patients don't know if they are getting the placebo or the actual agent. Further, the doctors who are doing the study shouldn't even know which patient is getting which drug or placebo. Subconscious unwitting bias can also be infectious.

Dr. Steed said the fact that three of the 12 patients in his study experienced complete wound healing after treatment by the placebo conveys a major lesson for nurses, physicians and others who work with patients.

The lesson is, "You make a difference."

Actually, that may be more than a medical lesson.

It may be a fundamental truth about human relationships in general.

Having Hiccups Funny, Annoying, Rarely Harmful
MAY 18, 1995

A James Madison University student attended a fund-raising auction not long ago while she was in the midst of a prolonged attack of the hiccups. She hiccuped during the bidding, and the auctioneer accepted her involuntary utterance as a bid. But the deed was undone, according to an article in *The Breeze*, the JMU student newspaper, when the auctioneer realized that the young woman was having the hiccups.

For the most part, hiccups are harmless and, at times, the object of some light humor. (Old joke: If buttercups are yellow, what color are hiccups? Answer: Burple.)

It's not clear to medical scientists why we hiccup.

"Hiccups," wrote Dr. Paul Rousseau recently in the *Southern Medical Journal*, "are a common annoyance serving no known physiologic function."

They occur when the diaphragm—the muscle that helps us breathe—contracts involuntarily, followed by a quick closing of the vocal cords. It's the rapid closing of the vocal cords that results in the typical sound of a hiccup.

Rousseau, who is with the geriatrics department at the Veterans Affairs Medical Center in Phoenix and Arizona State University's adult development and aging department, noted that the medical term for hiccups is singultus (sing-GUL-tus).

He said the term comes from a Latin word that means a sob or speech broken by sobs, or the act of catching one's breath while sobbing.

Most episodes of hiccups are harmless and short-lived, but in some cases they may last for days to weeks, sometimes with serious consequences. If someone has hiccups for more than 48 hours, Rousseau said, they should seek medical advice.

Rousseau and others say the benign, nuisance forms of the hiccups can have various causes, including excess consumption of food or alcohol, swallowed air, drinking carbonated beverages, sudden excitement or emotional stress. Rapid temperature changes, such as taking a cold shower, may induce a spell of hiccups; so can drinking hot or cold beverages.

If a case of hiccups lasts more than two days but less than a month, it's called persistent hiccups, and if it lasts longer, it's known as intractable hiccups. Rousseau said there are dozens of causes of persistent and intractable hiccups, and the adverse effects can include fatigue and exhaustion, weight loss, insomnia and, occasionally, death.

Treatments for the benign form of hiccups are perhaps the subject of more folk remedies than any other common malady. And the JMU student probably heard most of them, according to the report in *The Breeze*, including drinking water while upside down.

Rousseau cited such anecdotal hiccup cures as drinking water from the far side of the glass; biting on a lemon and swallowing granulated sugar.*The Merck Manual*, a medical handbook, gives credence to two well-known treatments: holding one's breath and breathing into a paper bag.

A high level of carbon dioxide in the blood inhibits hiccups; a low level accentuates them. Holding the breath or breathing deeply in a paper bag ("Not a plastic bag, as it may cling to the nostrils," warned the manual) each helps raise the level of carbon dioxide in the blood.

Physicians often try various drugs to treat stubborn cases of hiccups. In some severe, prolonged cases of hiccups, physicians may try, if all else fails, injecting a tiny amount of an anesthetic to the phrenic nerve, which controls the diaphragm.

But the first approach, Rousseau suggested, should be trying a home remedy.

And, he might well advise, stay away from auctions in the meantime.

Eating Insects Comes Natural to All Of Us
JUNE 15, 1995

Many people probably would consider pilot Scott O'Grady's act of eating ants to be one of the bravest things he did during his six days of hiding in a Bosnian forest.

In the Western world anyway, eating insects is generally considered to be an unappetizing thing to do, if not bizarre and repulsive.

But in fact, insects have been part of human fare for as long as there have been humans, perhaps a carryover from our prehuman ancestors' diets of termites, ants and other such creatures.

Images from ancient Egypt, dating to around 700 B.C., show that skewered locusts were delicacies at royal feasts. The Old Testament talks about eating locusts. So does the Mishnah Torah and the Koran, noted May R. Berenbaum in her book, *Bugs in the System.*

The Greeks of ancient times were insect eaters, she said, consuming such creatures as cicadas and grasshoppers. So were the Romans, whose insect delicacies included a grub or caterpillar "of uncertain identity," Berenbaum said.

Insects such as termites, locusts, ants, grasshoppers, cicadas, beetles and bees continue to be eaten by members of societies around the world, including some in Africa, the South Pacific, Asia and Australia. In Thailand, giant water bugs are for sale in open trays in marketplaces.

In the Southwestern United States, native Americans have sought out an insect called the honey pot ant. Those creatures store honey in their distended abdomens, according to Dr. Richard D. Fell, a Virginia Tech entomologist, and eating them "is like eating sweet globs of honey."

Fell and others note that, as a food, insects have a lot going for them. They are rich in fat and protein, especially protein. Fell said the protein content of live insects can range from about 20 percent to 60 percent, depending on the insect and which stage of its life cycle it is in. (For comparison's sake, a hamburger's protein content may range from 20 to 30 percent.)

The calorie content of insects, Fell said, is about 300 to 500 per 100 grams (between three and four ounces). Berenbaum said that a pound of grasshoppers contains about 1,365 calories, compared to 1,240 calories per pound for beef.

Survival experts have long pointed out the advantages of eating insects. Bradford Angier, in his classic book, *How to Stay Alive in the Woods*, quotes one experienced outdoorsman as saying that "insects are wonderful food, being mostly fat, and far more strengthening than either fish or meat. It does not take many insects to keep you fit. Do not be squeamish about eating insects, as it is entirely uncalled for."

The outdoorsman said nearly all insects encountered in the woods, including moths and may flies, "are very palatable."

At least one exception, according to Dr. Richard Mills, an entomologist at Virginia Commonwealth University, may be the cockroach. "They are real greasy," he said. Mills noted that insects' exoskeletons, their hard outer shells, are indigestible. But frying or roasting them, or catching them shortly after they have molted, helps improve their edibility.

Both Mills and Fell said they have sampled some insects. Mills, for example, said he tasted some of the grasshoppers that South Pacific Islanders roast and value as a delicacy. He reported that they are "very good."

Fell said he even prepares bee pupae—a pupa is a stage in an insect's development—in some of his classes. "We fry them right up right there," he said, adding that most students are willing to try them. But he said bee pupae don't have much taste.

If he fries them with a little garlic and a little butter, he said, the pupae taste like "a little garlic and a little butter."

THE WORLD AROUND US

No matter where you go, there you are.
From: *The Adventures of Buckaroo Bonzai*

Inspiration for columns comes from many sources. In some cases, such as "Digging To China Only Possible From Argentina," ideas are spinoffs from earlier interviews. Meandering rivers caught my attention during an airplane flight to a science meeting on the West Coast.

And when I'm on an airliner, I often keep myself approximately oriented by using my wristwatch and the sun as direction finders—a trick I once shared with readers. Column inspirations have also come from assignments to interesting, faraway places as well as from vacation trips.

Digging To China Only Possible From Argentina
AUGUST 18, 1991

It's one of those little myths we learned in childhood: If we were to dig a hole straight through the Earth, we'd come out someplace in China.

It's a myth that's been perpetuated in many ways in the lighter and fictional side of our culture, from Bugs Bunny cartoons to movie titles.

A few years ago, for example, the title of a serious, popular movie was *The China Syndrome*, based on the somewhat flip observation that if a nuclear reactor's core became uncontrollably active, it would turn into a molten, high-temperature mass that would melt right on down through the Earth, emerging in China.

The fact is, if we were to dig straight through the Earth from anywhere in the continental United States, we'd come out someplace in the Indian Ocean. If we were to dig such an impossible tunnel from anywhere in Virginia, we'd come out in that ocean several hundred miles southwest of Australia.

The train of thought leading to such musings was started a few weeks ago through an interview with Dr. Athelstan Spilhaus, a noted oceanographer and figure in American science who now lives near Middleburg. One of his interests for decades is making maps of the Earth that emphasize the oceans rather than the lands because, as he says, our planet really is a water planet, with one big ocean covering about three-quarters of it.

So his maps cut some land areas instead of the oceans. Or he uses natural boundaries, such as continental shorelines or tectonic plate boundaries, as edges for his maps, rather than arbitrary cuts down the middle of, say, the Pacific Ocean.

In the process of developing such maps, he's had to consider the tunneling-through-the-Earth problem.

As he points out in his writings, which include a new atlas of his maps and some articles in the magazine *Smithsonian* a few years ago, if you play the tunneling game on almost any continent in the world, you're likely to come out in an ocean and not on dry land.

Only about 3 percent of the Earth's land, he notes, is directly opposite other land. Greenland, for example, is opposite part of Antarctica.

But South America is the prime continent for land-to-land tunneling. Sumatra and Borneo are opposite the upper part of South America, around the equator.

And the lower part of South America is directly opposite a part of China.

So, if you really want to dig through to China, your chances are best if you begin the tunnel in southern Argentina.

Actually, the matter of tunneling to opposite points on the Earth has a more formal air to it in the map-making field. In cartography, pairs of opposite points are called antipodes, a term that goes back to Greek meaning opposite feet. The North and South poles are examples of antipodes.

But any pair of places that are directly opposite one another are antipodes. In her delightful book, *The Trigonometric Travelogue* (Pico Beach Books), Mimi Gerstell says that a group of islands south of New Zealand were named the Antipode Islands by British explorers because they are about opposite of London.

She, too, notes that most people on Earth would come up in an ocean if they dug holes through the center of the Earth. "Most North Americans would come up in the Indian Ocean; most Australians in the Atlantic; most everybody else in the Pacific."

Ms. Gerstell, incidentally, gives the rule for finding the point that's opposite of any given place. Pairs of antipodes have the same number for latitude, except one is north and the other south. And, their longitudes add up to 180 degrees.

The latitude of Richmond is about 37 degrees north and its longitude is 77 degrees west. The place directly opposite it is about 37 degrees south of the equator and 103 degrees east.

So Richmonders could place an X at that remote point in the Indian Ocean and declare it the Antipode Waters.

By the same token, seamen traveling in those waters could legitimately refer to Richmond as the Antipode City on their charts of the world.

Straight Answers Not Easy on Why Rivers Curve
SEPTEMBER 4, 1988

There's nothing like going into the third dimension (up) to get a fresh perspective on things that aren't readily apparent to us in our day-to-day, surface-clinging existence. A good way to enter the third dimension is to be an airline passenger with a window seat in a plane that's several miles high, as was my case during a recent weekend.

One of the intriguing sights from such a vantage point is that of the snake-like, undulating curved paths that rivers take. Rivers, especially older, mature rivers, do not travel in straight lines, at least not for long. Rather, they meander across the landscape.

Actually, one doesn't have to be an airline passenger to be aware of the sinuous nature of river courses. It can be seen on maps. And in some cases, it may be a regional natural attraction, like the seven bends of the Shenandoah River's North Fork, as seen from a Massanutten Mountain overlook east of Woodstock, Va.

But it takes an airplane flight over several hundred miles and over a number of rivers to give one a full appreciation of the fact that rivers just don't seem to like straight lines. Rather, they seem to be guided by a geometry that says the shortest distance between Point A and Point B on a plane is a curve.

Just why rivers do that is not a trivial problem. One of the 20th century's greatest physicists, Albert Einstein, was fascinated with the problem of meandering rivers and even wrote a scientific paper about it, which was published in 1926 in a German science journal.

Dr. Elizabeth A. Wood, a former research scientist with Bell

Telephone Laboratories, includes a discussion of meandering rivers (and a footnote about Einstein's paper) in her classic book, *Science for the Airplane Passenger*.

During the 1960s, two U.S. Geological Survey hydrologists conducted research on meandering rivers, and published a general account of the phenomenon in the June 1966 issue of *Scientific American*. From such sources come a picture of why rivers bend in a series of curves.

The meanders of a river are perpetuated by the erosive powers of flowing water. When the river curves, the water on the outside of the curve flows faster than the water on the inside of the curve. The faster-flowing water on the outside cuts into the outside bank, maintaining and even deepening the bank's concave shape.

The material eroded from the outside concave bank, meanwhile, is carried downstream where it is deposited on the inside bank. The pattern continues: The outside bank is progressively carved out and the downstream inside bank is gradually filled in.

A net result is that the meanders shift laterally over time. And it is the sideways shifting of the meanders that eventually produce the river's flood plain, which is a broad valley floor surrounding the river.

It's not floods that produce the flood plain but the changing snake-like pattern of the river itself. The flood plain is called that because it is the area that's flooded when the river overflows because of heavy rains.

Sometimes a river's waters will break across the neck of a loop, rather than traveling all the way around the bend. Such isolated segments are called oxbow lakes, and according to one geology textbook, the lower Mississippi abounds with oxbow lakes.

The question is, how do such meanders get started in the first place? Nature provides many opportunities for the courses of rivers to become diverted, such as fallen trees, large rocks and varying types of soils.

But, according to the U.S. Geological Survey hydrologists who studied river meanders, such irregularities aren't really necessary for meandering to occur. For one thing, random irregularities like fallen trees and locations of boulders cannot account for the generally regular meandering patterns that rivers actually tend to follow, according to the hydrologists, Luna B. Leopold and W.B. Langbein.

It turns out, they said in their 1966 *Scientific American* article, that meandering is a result of a variety of subtle chance events, like the turbulence of the water's flow, how the flow interacts with the river bed and how the river bed and the banks were formed in the first place.

"It is a paradox of nature that such random processes can produce regular forms," they noted. Leopold and Langbein said that meanders will generally appear wherever a river flows through a gentle slope, in a material that's fine-grained enough to be eroded and transported, but cohesive enough to make definite firm banks.

It also turns out, they said, that sinuous curves—regular patterns of undulating curves—require the least expenditure of work by flowing water that is forced to make bends and turns.

Dr. Wood, in her book, refers to meandering rivers as being mature rivers. In them as with many other facets of life and nature, it may be a sign of maturity when a way has been found to do something with the least effort.

In the case of rivers, the curves of least effort are especially pleasing to the eye, from a few thousand feet to a few miles up.

Wristwatches Can Fulfill Dual Roles
AUGUST 9, 1993

A few years ago, I bought a digital wristwatch, not so much for its digital display as for the compass that was in the wrist band. Every once in a while I am curious about directions, like the heading of the airliner I'm in or the orientation of the streets or roads in places that are new to me.

Digital wristwatches may be fine for giving the time in a user-friendly format—they show that it's 4:57 without having to count small marks on a dial. But they don't have the intimate relationship with the solar system that traditional watches have.

Because of that relationship, the traditional hands-and-dial watch is a reasonably good direction finder in itself—when it's daytime and the sun is visible. Before I bought the watch with the wristband compass, I often oriented myself with my old-fashioned hands-and-dial watch.

The rule is this: Point the hour hand toward the sun. South lies halfway between the sun-pointing hour hand and the 12 on the watch dial.

Now that the original compass-holding wristband on my digital watch has been replaced, I no longer have the compass that persuaded me to go digital in the first place.

And it's almost impossible, unless you have a vivid imagination, to use a digital watch as a compass—even if it is daylight and the sun is shining.

The sun is the key to why the traditional hands-and-dial watch makes an approximate compass; it is the link between such timepieces and the solar system.

First of all, the sun moves across the sky at a rate of about 15 degrees per hour. Actually, it's the spin of the Earth that makes the sun appear to move across the sky. The Earth makes one spin in 24 hours. One spin means that any given point traces out a circle, 360 degrees; 360 divided by 24 gives 15 degrees per hour.

The hour hand on a watch or clock, meanwhile, moves at a rate that's twice that of the sun's pace across the sky. There are 12 major divisions on the clock's face, so a trip around the face is 360 degrees divided by 12. That means the angular distance between each major division—between 12 and 1, 1 and 2, 2 and 3 and so on—is 30 degrees. So, in one hour the sun moves 15 degrees and the hour hand of a watch moves 30 degrees.

One other fact is needed to explain the watch-as-a-compass trick, and that's the definition of noon as reckoned by sun time. Noon is when the sun is on an imaginary arc in the sky that runs precisely north and south. That arc is known as the meridian.

That's the M-word in the abbreviations a.m. and p.m. A.M. stands for ante meridian, meaning that the sun has not yet crossed the meridian, and p.m. stands for post meridian, meaning that the sun has passed the meridian. When the sun is on the meridian, it is directly south.

So if you pointed the hour hand of a watch toward the sun at noon (or 1 p.m. when Eastern Daylight Time applies), both the hand and the number 12 on the dial would be aligned with the sun, and they would be pointing due south.

If you pointed the hour hand toward the sun at 2 p.m., the sun is 30 degrees west of the meridian, or 30 degrees westward from due south. The hour hand, however, moved 60 degrees during that period. So half the distance between 12 on the dial and the 2 is 30 degrees—and the direction of the meridian, or south.

The rule—south is halfway between 12 and the hour hand when the hour hand is pointed toward the sun—applies generally for all times during the day. (Again, remember that daylight-saving time is an hour ahead of sun time.)

It works because the traditional so-called analog watches and clocks are related in an intimate way with the motions of the sun across our sky. Digital clocks and watches, on the other hand, are linked to the oscillations of quartz crystals set vibrating by a battery.

So I may soon go back to the old-fashioned, hands-and-dial watch. Its innards may be quartz crystals and tiny circuit boards, but its hands and face will give me a connection once again with the grand motions of the outer world.

Besides, it can orient me in space as well as in time.

West Displays Evidence Nature Is Never Content
AUGUST 2, 1987

"Observe always that everything is the result of change, and get used to thinking that there is nothing Nature loves so well as to change existing forms and to make new ones like them," wrote Marcus Aurelius more than 17 centuries ago.

A vacation trip to the western part of the United States reinforces the Roman emperor's observation. From the Badlands of South Dakota to the majestic canyons of Utah and Arizona, the predominant theme of the landscape is change. The message is Marcus Aurelius's message: Nature is never content with the status quo; it is always adjusting, adapting, changing.

Thus, a vast shallow sea bed becomes, in time, an expansive flat prairie. And in time, the erosive forces of rain, wind and frost carve through the layers of soft rock and soil to create canyons and gullies and structures that descend hundreds of feet into the once tabletop-flat prairie, in the region we now call South Dakota.

The carvings, furthermore, reveal through colored horizontal layers the changes that occurred during the millions of years since the ancient sea covered the area. Near the tops of the rugged, crudely carved forms, for example, are layers of whitish material, the ashes of volcanoes that erupted farther to the west and southwest.

And volcanoes themselves were, and are, among nature's tools of change, as reflected by the still-desolate landscape around Sunset Crater in Arizona, where trees and other vegetation are only now beginning to gain a tentative footing in the black ashes and cinder fields of a volcano that erupted more than 900 years ago.

Even the volcanoes are subject to the other forces of change—wind, rain and ice. In northern Wyoming, there is a 900-foot-tall tower of rock, the core of an ancient volcano, that can be seen from miles around and which figured prominently in a popular movie a few years ago, *CloseEncounters of the Third Kind*.

Some 60 million years ago, molten rock forced its way up from deep within the Earth's interior to near the surface and cooled in crystalline form, producing column-like structures. Then, wind, rain and ice leveled the land around, leaving the core progressively exposed until it became a dramatic tower that today is called Devils Tower.

Even now, the changes are continuing. Gravity, the great leveler of mountains and other structures that dare stand above surrounding countryside, pulls unrelentingly on the tower's stone columns, some of which become loosened in time by the forces of freezing and thawing.

So, once in a while, an occasional chunk tumbles down, joining a slowly growing pile of stony rubble at the tower's base.

The evidence of nature's propensity for change exists elsewhere around the country, of course, but in the West and Southwest it is particularly obvious to even the untrained eye. The scenery there brings home the notion that our planet is still alive, that its inner fires are still active and that its surface is always changing.

Exploration by unmanned spacecraft of the other planets in our solar system, particularly the inner ones, has provided some perspective. The inner planets, from Mercury, close to the sun, to Mars, the next one out beyond the Earth, are rocky planets; the outer planets are composed primarily of gases.

Except for the Earth, the other rocky planets appear to be dead. Their inner fires have cooled, and with their cooling has come unchanging sameness. All of the inner planets, according to current thinking, were formed with hot interiors, from which hot molten material and gases spewed, keeping the planets' surfaces and atmospheres renewed and changing.

But then, the fires dwindled and the surfaces cooled. It happened first with tiny Mercury, with our moon and then with Mars. (The nature of the surface of Venus, obscured by thick clouds, is still being debated by scientists.)

Volcanoes became quiet; there were no longer new supplies of gases for the atmospheres, and gradually the atmospheres became thinner or vanished.

Only the Earth still lives. It is the one rocky planet that is still hot within, as vividly reflected by volcanic activity, by the shifting and movement of the continents and by the steamy thermal basins of Yellowstone National Park.

And it has large amounts of water, with a dynamic atmosphere that helps keep the average global temperature between water's freezing and boiling points.

The Apollo 8 astronauts—the first humans to travel far enough away from our planet to see it as a small ball in the vastness of black space—remarked that the Earth, as seen from the vicinity of the moon, was the only colorful object they could see. It was a serene blue, due to the ample supplies of water in the atmosphere and oceans.

And beneath the serene blue face that the Earth presents to space travelers is a planet whose surface continues to be shaped by forces that are directly and indirectly related to a hot interior that's keeping the whole thing alive.

NATURE— HUMAN

What a piece of work is a man!
WILLIAM SHAKESPEARE, *Hamlet*

One of the early dogma-breaking events that occurred during my years of covering science was the discovery that we and other creatures can influence our autonomic nervous systems—the systems that control such functions as heart rate and blood pressure. The implications of those pioneering studies, such as the one described here, are still with us, suggesting insights into psychosomatic illnesses and helping give respectibility to meditation as a means of coping with daily stresses.

The column on negative thoughts gave me an opportunity to draw a quotation from one of the first stories I helped cover for the Times-Dispatch—William Faulkner's first day as writer-in-residence at the University of Virginia on February 15, 1957.

The final colum in this section was one of my earliest, and one that because of its implications was the most enduringly haunting for me. The experiment I reported on indicated that ordinary Americans are potentially capable of inflicting pain on fellow citizens when ordered to do so under non-coercive circumstances, and the study has since become a classic in social psychology. It has also since raised ethical questions over the manner in which the experiment was conducted.

Responsibility and Biofeedback
DECEMBER 5, 1971

The motion picture showed a white rat, completely paralyzed by an injection of curare, fitted with a small face mask. The mask was connected to an artificial respirator, which breathed for the rat until the curare's effects wore off.

The heart was beating normally; the rat's brain was functioning. But otherwise, the animal was immobile.

The scene switched to an instrument that provided a digital readout of the rat's heart rate, in beats per minute. The numbers flashed quickly on the instrument's face—395...405...400...408...395...403. The normal rate, explained the movie's narrator, is around 400.

The filmed experiment was intended to show that the rat was capable of "willing" its heart to speed up. Whenever its rate rose above 405 beats per minute, a drop of sugar water would be released onto the animal's tongue, through the face mask.

Soon, the numbers flashing on the monitoring instrument rose—408...406...411...405...407...404...400....A continual tracing of the rat's heart rate on a paper tape showed a gradual rise.

When the rat's heart rate became consistently greater than 405, a new goal was set by the experimenters—410. And when that was uniformly surpassed by the animal, 415 beats per minute was set, and so on by increments of five.

An hour or so passed, represented symbolically by a change of scene on the movie screen. The broad paper tape was moving steadily under the fixed stylus, showing consistent heart rates in the 480-490 range. Someone folded the tape to show the contrast between the heart rates at the experiment's beginning and after a period of time. The distance between the two slightly jagged lines was remarkable; the immobile rat had apparently, indeed, "willed" its heart rate to increase steadily by nearly 100 beats per minute with the reinforcing reward of drops of sugar water.

The movie, in which the rat experiment scenes were only a segment, was shown to a group of visiting science writers at State University of New York Stony Brook. Afterward, Dr. Lester G. Fehmi, assistant professor of psychology at the university, told the visitors that the opposite experiment—slowing the heart rate—could also be done. In fact, there have been instances when the experimental rats slowed their heart rates to such a sluggish pace that the animals died.

Further, he said, researchers have been successfully able to encourage rats and other animals to increase or decrease the blood flow to some of their body structures. They could, for example, make their left ear blush while their right ear retained its original color, or vice versa.

These particular kinds of rat experiments, which have been conducted in one form or another in various laboratories for several years, have helped overthrow a dogma and usher in an enigma. The dogma has been the generally accepted belief that the autonomic nervous system—

the system that controls such functions as heart rate, blood pressure and the various activities of the viscera—is truly automatic and independent of the conscious control. You can raise your arm, flex a muscle or blink an eye whenever you wish, but you cannot increase blood flow through your kidneys or "turn off" the flow of gastric juices to the stomach at will, it was believed. The enigma is, how can such things be done—now that it has been demonstrated by various animal studies that they can?

Further, more recent studies with humans suggest we, too, can control our heartbeats, our blood pressure and other activities that we normally are now aware of. We can somehow permit one brain wave to predominate over others (practitioners of yogi can routinely do this); we can increase heart rates, decrease blood pressure and, suggested Dr. Fehmi, we may even be capable of discharging single nerve cells, one at a time in different parts of our bodies, at will.

The trick is knowing when you've influenced the function or activity in question. In the experiments conducted thus far by Dr. Fehmi and others, a tone is sounded or a light comes on when the intended response is achieved. Such an audible or visible signal of success seems to be a sufficient reward for humans, a more abstract equivalent of sugar water drops.

Interestingly, subjects who learn how to keep a tone sounding, or a light lit, by maintaining a give brain wave, or blood pressure, or heart rate, say they aren't sure how they do it. But once they've learned with the aid of feedback devices, they apparently can "turn on" the response almost anytime, anywhere.

At the moment, no one quite knows whether such techniques have a therapeutic value—whether, for example, they could be used to treat people with high blood pressure. The indications are that humans can learn to lower their blood pressure to at least some extent in a laboratory setting, but whether they can lower it sufficiently and consistently outside the laboratory remains to be seen.

But the present studies, as dramatically depicted by the movie scenes of the curarized white rat and the graphic displays of its rising heart rates, may have other kinds of value. They may help explain, for example, how psychosomatic illnesses come about. They may help explain how one girl in a survival training course survived a five-day outing in the wilderness while her two companions despaired and died. (The survivor in this incident said she was determined to survive, while her two companions just seemed to give up when the weather conditions became severe.)

In a more philosophical vein, the images of the immobilized rat and its apparent control over its heart beats may convey a rather provocative message:

We may be potentially more responsible for ourselves than we realize or even know how to be.

Thoughts Tend to the Negative
OCTOBER 18, 1992

During his first day as writer-in-residence at the University of Virginia in February 1957, William Faulkner was asked why his books contain so much violence and examples of evil and meanness.

"People are quicker to be interested in violence," Faulkner responded. "More people will stop to see a fight on the street than would stop to hear Christ."

Faulkner is not the first to be criticized for focusing on the negative side of humanity. The news media, particularly, are often scolded for playing up slayings, rapes and other dark things that humans do.

Callers and letter-writers complain that we feature the bad news and that we ignore the good things going on around us. In fact, most news organizations develop many, many articles and programs about the selfless deeds or helpful acts that individuals and groups carry out.

But, it seems, such articles are not remembered as much as the ones about the seamy side of humanity.

Is it possible, therefore, that we pay more attention to bad news than to good news, even though we proclaim a preference for good news? Do we have some sort of built-in antennae for the negative?

It is indeed possible, according to some studies led by a Stanford University psychologist.

The psychologist, Dr. Felicia Pratto, and Dr. Oliver P. John, a psychologist at the University of California at Berkeley, found through a series of experiments that people do tend to have an unconscious bias toward negative things. Some of their work has been published in the *Journal of Personality and Social Psychology.*

One series of their experiments required volunteers to name the colors, quickly, of colored letters that flashed on computer screens. The letters formed words that are generally regarded as having negative, positive or neutral meanings.

Curt, rude, wicked and lazy were among the words with negative connotations; stable, polite, loving and kind were among those on the positive words list. Neutral words included desk, floor and sidewalk.

The volunteers consistently, and unconsciously, dwelled fractions of a second longer on negative words than on positive or neutral ones, the psychologists found. The volunteers were Berkeley students; a total of 65 were involved in three sets of tests.

The students weren't aware that they were dwelling a split second longer on the words with negative meanings; in fact, the words were flashed on the screen so fast that they claimed that they really didn't see them. It happened so quickly that the words didn't register in their minds, they said.

But when they were pressed to try to recall them, Dr. Pratto said the volunteers consistently cited more negative words than positive ones.

So, negative things seem to have a particular attention-grabbing power for us, Dr. Pratto says.

And she says there may be a reason. We probably have such an automatic vigilance system—biased toward negative information—because it proved useful in alerting our ancestors. The system, she said, holds negative data for an extended instant, as if telling us to pay attention in case we need to respond.

Now that we have such insight into ourselves, can we alter the automatic vigilance system so that it doesn't focus so much on the negative? Dr. Pratto said research indicates that people cannot stop the system, even when they're aware of it.

But, she stated in a news release from Stanford that researchers don't know yet how the vigilance mechanism ties in with other thinking processes. If such connections are found, perhaps ways of altering the mechanism may be discovered.

We are not completely negative-oriented beings, incidentally. Dr. Pratto did find one category of words with positive connotations that commanded as much attention as did the negatively-tainted ones.

Those words, she reported, were ones related to reproduction, such as babies or sex.

Maintenance of Humaneness Breaks Down Under Orders
OCTOBER 4, 1964

Not long ago, a group of psychology majors at Yale University were asked how many people would continue to inflict severe pain upon another person—who had already been experiencing a great deal of pain —if ordered to do so.

The most pessimistic of the students said perhaps three out of 100 would follow the orders. Some said none; the others said maybe one or two out of 100 would continue inflicting the pain under the circumstances that were described to the students.

It turns out that the students (and some professional psychologists whose replies were similar to the students') were drastically wrong. They seriously overestimated the capacity of people to maintain their humaneness under orders, and a non-coercive set of orders at that

The experiment described to the students, and to the psychologists, was reported in the October, 1963, issue of the *Journal of Abnormal and Social Psychology* by Dr. Stanley Milgram, formerly at Yale, and now at Harvard University.

In the experiment, there was an "experimenter," a "victim" and a volunteer subject. For all the volunteer knew, the "victim" was just another volunteer.

The "experimenter," played by a 31-year-old high school biology teacher with an impassive manner and whose appearance was somewhat stern, told the subject and "victim" that they were about to take part in a learning experiment.

More specifically, they were told that the experiment was designed to study the effects of punishment on learning—whether punishment affected learning, and if so, how much punishment would be a factor, and so on.

Then, one was to play the role of teacher, and the other the learner. The choice of who was to play which role was determined by drawing slips of paper; the drawing was rigged so that the "victim" was always the learner.

The victim was taken into a room and strapped to an "electric chair." The teacher and volunteer and the experimenter then went into an adjoining room where there was an electrical panel board. On the board were 30 switches, labeled from 15 to 450 volts. Groups of the switches were labeled from "Slight Shock" to "Danger: Severe Shock."

The teacher was to give the learner a quiz on groups of words. Each time the learner made a mistake, the teacher was to throw one of the switches, and he was to give higher and higher voltages as the experiment progressed. The learner indicated his answers by pressing a button on the "electric chair," which in turn turned on a light in the control room. The learner, of course, was actually receiving no electric shocks. And, he would purposely miss three out of four questions put to him by the teacher. Further, when the teacher reached the "300-volt" switch, the learner would begin pounding on the wall with his feet.

Beyond that point, he would no longer respond to the questions by pressing the button on his chair. The teacher could not see the learner, but he could hear the violent kicking.

If the teacher hesitated to throw the next higher switch after hearing the pounding, the experimenter would use a series of verbal prods, like, "Please continue," "The experiment requires that you continue," "It is absolutely essential that you continue" or "You have no other choice, you must go on." The prods were used in that order.

This experiment was performed with 40 volunteer subjects, whose ages ranged from 20 to 50. Their educational levels ranged from one who had not finished elementary school to those who held doctorates and other professional degrees.

Twenty-six of the 40, reported Dr. Milgram, obeyed to the end, giving the maximum of "450 volts," which was two steps beyond the verbal designation on the switches that said, "Danger: Severe Shock." This was 150 volts beyond the point where the "victim" began pounding on the wall, and where he ceased pressing his light button.

Five refused to go beyond the 300-volt level; four gave one further, higher shock, then refused to continue; two broke of at the 330-volt level, and one each at the 345, 360 and 375 volt levels.

"Upon command of the experimenter, each of the 40 subjects went beyond the expected break-off point," Dr. Milgram reported. None stopped prior to the 300-volt level, when the learner began kicking.

And, although the obedient subjects continued to administer the "shocks," they often did so under extreme stress. Several experienced "full-blown, uncontrollable seizures"; a number developed nervous laughing fits, sweating, trembling, groaning, biting of lips and digging fingernails into their flesh were observed among a number of others, Dr. Milgram said.

Hence the psychology majors were entirely too optimistic in their guesses of how people would react in this experiment. It appears that obedience to orders delivered in a "firm, but not impolite" tone of voice can override training in ethics, sympathy for fellow humans and moral conduct.

At one point in his paper, Dr. Milgram quotes the following remark by the noted English author, C. P. Snow:

"When you think of the long and gloomy history of man, you will find more hideous crimes have been committed in the name of obedience than have ever been committed in the name of rebellion."

Snow was specifically referring to the German officers in the Third Reich.

TRIVAL MATTERS

What mighty contests arise from trivial things.
ALEXANDER POPE, *The Rape of the Lock*

Science is a rich source of trivial facts, and to a trivia lover like myself, some odd facts stick with me from my wanderings through journals and books as cockleburs stick to a hiker's clothing during a walk through a shrub-filled woods. Sometimes, a bit of trivia became inspirations for columns, as the following collection demonstrates.

Memorized Code Sentences Ease Learning
SEPTEMBER 1, 1994

The late Richard Feynman, the physicist who became known to the public through popular books by and about him, once dazzled some mathematicians by his ability to solve complicated problems in his head.

Actually, he disclosed in the book, *Surely You're Joking, Mr. Feynman*, he wasn't the genius the mathematicians thought he was.

He had memorized a few key things in math, such as a particular logarithm, that he often needed in his physics calculations. And as luck would have it, the problems the mathematicians gave him could be solved by using the relatively few things he had memorized.

Few people can match the dazzling mental feats of a Feynman. But many of us can give the appearance of having slightly more mental power than we really do by remembering just a few key things, or words, or sentences. It's a trick that many a student has used, especially in the sciences.

For example, the names of the planets in the order of their distances from the sun can be recalled by remembering the following sentence: "My very educated mother just served us nine pizzas." The first letter of each word stands for a planet: Mercury, Venus, Earth, Mars, Jupiter, Saturn, Uranus, Neptune and Pluto.

Math students may turn to "My dear Aunt Sally" to guide them through a long equation involving multiplication, dividing, adding and subtracting. According to "My dear Aunt Sally," you do the multiplying first, followed by the dividing, then the adding and finally the subtracting as you come to each complicated term in the equation..

If you want to remember the first 15 numbers of pi, recall the following sentence: "How I want a drink, alcoholic of course, after the heavy lectures involving quantum mechanics." The numbers of letters in each word are 3.14159265358979.

Medical students have many such mnemonic aids to help them. A classic is, "On old Olympus' towering tops, a Finn and German vended (or viewed) some hops," which, by the first letter of each word, helps anatomy students recall the proper order of the cranial nerves, the dozen paired nerves that emerge from the brain. They are olfactory, optic, oculomotor, trochlear, trigeminal, abducens, facial, auditory, glossopharyngeal, vagus, spinal accessory and hypoglossal.

And the general function of each of those nerves—whether it's sensory, motor or both—can be remembered, in order, by the sentence, "Some say marry money, but my brothers say bad business marry money." Words starting with an S refer to sensory nerves; the M words stand for motor nerves, and the B words denote both sensory and motor functions.

Another is, "Ten zebras bite my candy." That, presumably, helps the student remember the branches of the facial nerve (temporal, zygomaticofacial, buccal, mandibular and cervical). The bones of the wrist, meanwhile, are supposed to be recalled by remembering, "Never love Tilly, Papa might come home." (The key: navicular, lunate, triquetrum, pisiform, multangular, capitate and hamate.)

For evaluating a patient's mental status, there's the sentence, "A sick man always thinks oddly," which reminds students to consider appearance, speech, memory, affect, thought processes and orientation.

Now, if you can just remember all of those sentences and what they stand for, and if someone should just ever ask you the right questions, you, too, might give the impression of being dazzlingly brilliant.

Actually, you would be if you could recall all those sentences and what they stand for.

Candidates for 'Wonders of World' List
JANUARY 11, 1987

A few years ago, the physician-essayist Lewis Thomas drew up a list of his candidates for seven wonders of the modern world, to replace what he called "the old biodegradable Wonders, the Hanging Gardens of Babylon and all the rest."

It wasn't his initial idea to develop such a list, he wrote in the June 7, 1983 edition of the *New York Times*; it was a magazine editor's. And it was only after considerable brooding over the subject that Dr. Lewis compiled his list, which included a beetle that lives exclusively in mimosa trees, a particular virus, the cells that enable us to detect scents and odors, termites, a human child and the living, dynamic planet Earth.

There's something about such lists that makes one consider drawing up one's own. Lists of wonders are generally cast in several ways. The traditional list of the seven wonders of the ancient world, for example, contained man-made structures, such as the Egyptian pyramids, the 400-foot-tall Pharos lighthouse, the Colossus of Rhodes and, indeed, the biodegradable Hanging Gardens of Babylon.

My own list also includes human creations from the world of technology. Rather than the exotic, it tends to focus on objects and devices that are so much with us that we tend to overlook the magic and marvel involved with them. It is a list of items that someone from the past, one or two centuries ago, would consider truly wondrous if they could somehow visit our time and culture.

Candidate number one would be color television. The worth of many of the programs that appear on television may be debated, but the technical aspects involved in sending and receiving color images, in motion and with sound, are impressive.

Consider the task. A scene, a play, an event is converted to electrical impulses, which are made to radiate as unseen weak signals through space. In many cases, the signals may literally travel into space to a satellite that re-beams them to earthbound stations.

The tiny signals then arrive at our home set, which detects them, amplifies them, turns them into color images and endows the images with sound. The device that does all of that is one of the most complex electronic items in our homes.

Candidate number two is the airplane. Less than a century ago, it was impossible for even one person to fly in a heavier-than-air craft. Now, airliners as large as a couple of railroad cars daily carry thousands of people thousands of miles within hours. A single airliner is larger and holds more people than the early ships that brought colonists to America.

What's so wondrous about the airplane is how some relatively simple principles of physics can be applied to make tons of metal and cargo fly though the tenuous atmosphere.

Candidate number three is also a vehicle of transportation, namely, the automobile. It is a mass-produced, complex machine that, on the whole, is a reliable, comfortable means of getting from point A to point B. Particularly wondrous about the automobile is its engine, an exquisitely well-timed, balanced and durable device that smoothly turns gasoline explostions into motion.

Candidate number four is the general power of modern medicine, from surgery to drugs, from artificial body parts to the ability of physicians to look inside the body, to see what's wrong, without surgery. In only the past few decades has medicine been able to control reasonably well many of the ailments that were the major killers of people throughout most of recorded history, such as pneumonia, tuberculosis and infections.

Candidate number five would certainly be the computer, which has permeated late 20th century life in the United States so thoroughly that daily events and transactions would be seriously hampered, if not halted, should all computers be simultaneously disabled. It's not just computers by themselves that are so wondrous, but the seemingly unlimited things that people can make them do, from word processing to diagnosing diseases, from guiding spacecraft to other planets to running factories.

Candidate number six is magnetic recording tape, an inexpensive, flexible and compact medium for capturing sounds and sights with good fidelity. Motion pictures that require a couple of large reels of celluloid film are now being put onto magnetic tape in a package about the size of a medium-sized book. Entire symphonies, or plays or language instruction can be placed on cassettes that can fit into a shirt pocket.

Candidate number seven would be all of the things that scientists and engineers are now able to do with products made from common sand. From it comes the raw material for the chips and related components that now make possible most of the other modern wonders, such as our color TV sets, radios and navigation equipment aboard airliners, the electronic ignitions of our autos, the computers that are now an integral part of all our activities, including medicine, and the machines that play our magnetic tapes.

In the end, that may be the greatest wonder of them all. We have learned to turn one of the Earth's most common materials into some of our most wondrous products.

Falling Corn? No Mundane Answers, Please
SEPTEMBER 28, 1986

A National Public Radio news program last weekend featured a woman from the Midwest who claimed corn kernels occasionally fell on her property and in her neighborhood from the sky.

She hoped, she told the interviewer, that scientists would seriously study the phenomenon, which had been puzzling her and her neighbors for some time. She also hoped, she said, that the explanation of the raining corn would not turn out to be something mundanely rational.

Her attitude is not unusual. Many people want to believe in strange and wonderful things that lie outside the rational realm of scientific explanation. If science can explain it, then it's dull, humdrum and ordinary; scientific explanations steal the romance, magic and mystery from events, according to them.

As a result, there is considerable appeal for such topics as UFOs, visits now and in the ancient past by extraterrestrial creatures, poltergeists and the Bermuda Triangle.

Now if corn is falling occasionally in the Midwest woman's neighborhood, the chances are that the explanation will turn out to be a rational one. Her plight is not entirely without precedent. There have been other instances in which strange, unlikely objects reportedly fell from clear skies.

In the 18th century, French farmers (and probably others elsewhere around the world) repeatedly told scientists that rocks occasionally fell from the sky. For quite a while, scientists refused to believe that rocks could fall from clear skies, then they came to realize the existence of meteors, which are chunks of rocky debris floating around in our solar system.

Now if anything, the discovery of meteorites has enhanced our sense of mystery rather than taken away from it. They have become scientists' direct links with the cosmos. They are telling them something about the nature of extraterrestrial materials; some of them may contain clues about chemical processes leading to life.

A larger message is that the rational stories that scientists develop through their studies generally enhance our sense of wonder rather than take away from it. In a sense, there is still magic in the world, even though there are scientific, rational explanations.

Take, for example, the matter of how new life arises and grows. Put a fertile seed into the ground and soon elements and molecules are organized into a plant or tree whose growth depends upon the orchestration of light, water and soil nutrients.

A seed is really an agent of organization; it contains instructions on how various elements should become organized to produce a variation of its parents. In creatures like us, the equivalent of the seed is a fertilized ovum, and the elements it organizes are nutrients supplied through the mother's body.

Scientists understand much of how seeds and their counterparts in the animal world carry out their organizing tasks, and they understand how the instructions for passing along new life and for growth are written in a chemical alphabet.

But understanding it doesn't take away the marvel of the process, the exquisiteness of the plan and how it is implemented. There is a rational scheme behind it all, but it nevertheless is awesome.

The scientists' stories about origins of another sort, namely the origin of our world, are also rooted in rationalism, but they also are awe-evoking.

For them, the story starts coming into focus a few trillion trillion trillionths of a second after the moment of creation, and within the first few minutes after it all began, they say that the basic ingredients of all material things—the particles of which atoms are made—came into existence from pure energy.

Over the ensuing eons, stars and galaxies developed from simple atoms, and inside the stars, many of the elements necessary to our existence were cooked, including the iron that carries the oxygen around in our blood streams.

The stars eventually died violent, explosive deaths; their cook pots of elements were shattered and the elements were spewed out, to mix with the dust and gas of the space between the stars. In time, that dust and gas and other material, drawn slowly together by gravitation, began swirling and collapsing to form a new star with planets.

And on at least one of those planets emerged life, and from life emerged creatures that tried to figure it all out.

And among those creatures were some who wondered why corn should fall from their skies, and hoped that the explanation would be more wonderful than any that rational scientists could spin.

Spontaneous Combustion
AUGUST 22, 1982

A few weeks ago, a spectacular story about a so-called case of spontaneous human combustion in Chicago appeared in newspapers and on television newscasts throughout the nation.

There was something medieval about the story, something suggestive of a rumor of witchcraft or supernatural demonic act in a village far, far away. But here it was, in Chicago, in 1982. A witness claimed he saw a woman a block or more away walking across a street and when he again saw her, she was aflame.

The body was burned extremely badly. And despite denials by a coroner that spontaneous human combustion can occur, the stories flowed anyway, claiming that eight or nine other such cases were recorded within recent centuries.

One wire service's citations of the cases were obviously drawn from *The Book of Lists*, in the same section that includes a list of "radiation" measurements of the brains of Leonardo da Vinci, Edgar Allan Poe, Billy Graham and 57 other people; 10 eyewitness accounts of levitation; nine possible visitations from outer space and "10 ghastly ghosts."

A day later, a Chicago coroner announced that the body had been dead for about 12 hours before the fire occurred. While someone who apparently collects reports of such cases continued to maintain the reality of the phenomenon in other instances, the whole bizarre, sensational episode of spontaneous human combustion in Chicago quickly extinguished itself. We returned from the Middle Ages.

Meanwhile, for one day, the country was fascinated by the possibility that the human body could spontaneously ignite, while walking down the street, or (according to *The Book of Lists*) while sipping tea or waltzing in a dance hall.

Skepticism was relaxed. After all, it is far more wonderful to believe that the human body could flare up (during a heated argument, perhaps?) than to consider how the body could become so hot in the first place. The body is a rather watery thing; it's around 60 percent water. So it would seem that the ignition of a body would require a tremendous spontaneously developed temperature.

I'm not sure how high a temperature would be required, but the ignition temperature of gasoline is between 500 and 800 degrees Fahrenheit; of charcoal, around 650 degrees, and of various types of coal, from about 750 to more than 1,000 degrees. The human body's normal temperature, meanwhile, is slightly less than 100 degrees Fahrenheit, and temperatures just a few degrees greater than 100 can be life threatening.

That's not to say, however, that there's no such thing as spontaneous combustion. As many farm boys know, at least of my generation, occasional summertime barn fires have been due to the spontaneous combustion of fresh hay that had been stored in them. A survey of barn fires in Wisconsin some years ago indicated that nearly 20 percent of the total damages were due to fires started by spontaneous combustion.

But there's no comparison between what can happen inside barns and what can happen inside the human body. (Actually, some of the cases of so-called human combustion in *The Book of Lists* allege the spontaneous ignition of an exterior part, like a leg or arm.) The spontaneous combustion of a plant material, like hay, or of other material, like a pile of rags saturated with paint or turpentine, requires several special conditions. One is that the material be stored in bulk so that the heat that builds up cannot be dissipated effectively.

In the case of plant material, the initial heating may be due to the normal respiration of the freshly cut plants. One product of respiration is heat. Further heating comes from the respiration of bacteria and fungi that thrive in relatively high temperatures, around 140 or 150 degrees Fahrenheit.

By the time such temperatures are attained, another heating mechanism has begun. It is due to the combining of oxygen from the air with the material. That process becomes the dominant source of heating above 165 degrees.

Because of the bulk, the heat accumulates. Because the heat accumulates, the material becomes hotter. The hotter it becomes, the greater is the rate of oxidation, which makes things even hotter until eventually, the ignition point of the material may be reached.

With materials like hay, according to a 1973 University of

Wisconsin newsletter's account of spontaneous combustion, several conditions must be met before ignition temperatures can be reached, and fortunately, all of those conditions are only occasionally fulfilled. The heat must be generated faster than it is dissipated; there must be enough oxygen and a certain amount of moisture must be present.

A critical situation exists when temperatures reach around 190 degrees Fahrenheit, according to the newsletter. At that point, there is an imminent danger of ignition of the hay.

So there is such a thing as spontaneous combustion.

And to many a farm boy of an era past, the occasional barn set afire by spontaneous combustion was a magical, mysterious thing. Even when the facts are known, the whole process is still wondrous to contemplate. Materials can consume themselves with a fiery flare under certain circumstances, due to natural processes.

But such processes are not applicable to the human body. Even if they were, the body's delicate and vital molecules would be destroyed by heat long before its ignition temperature would be reached.

"Burn-out," "fiery temperament" and becoming "hot under the collar" may be good figurative terms, but in the literal, scientific sense, they're not so hot.

FUN AND GAMES

A Briefer History of Time—BANG!
MICHELE M. MEAGHER
Sex as a Heap of Malfunctioning Rubble

I have a special fondness for the kind of science that makes us smile, or exclaim, "How about that!," or directs our attention to the complexity that can underlie some of our everyday experiences. Crushing a mint Lifesaver with a pair of pliers in a dark room generates sparks, and scientists are still trying to figure out the precise reasons. A folk craft device seems magical, and physicists debate how it really works. Popular movies, especially the space fantasy ones, are great vehicles for scientific analyses—of principles violated.

One of my favorite, early "fun-science" pieces is the report on an actual study, reported in Science, *that found an amazing similarity between the then newly arrived moon rocks and cheeses. One of my favorite, recent "fun-science" pieces is the one about how the universe's basic laws determine that toast is almost always bound to hit the floor jelly-side down.*

Then, of course, there's the analysis of walking vs. running in the rain that concludes: "Maximizing the velocity through the rain will be rainstrike minimizing."

Lifesavers Flash a Bit of Scientific Enlightenment
JUNE 19, 1988

It was an observation that, you might say, struck her in a flash.

Why, wondered Linda M. Sweeting, then a chemistry graduate student, did those crystals that she scraped off glassware following her experiments give off tiny lightning-like emissions during the scraping process?

About 20 years later, high technology and wintergreen Lifesavers are helping her look more deeply into the phenomenon.

In fact, Dr. Sweeting, now a professor of chemistry at University Towson State University in Baltimore, has studied the puzzle to a degree of sophistication that's only been possible within the past few years. She reported some of her results a week ago at the 3rd Chemical Congress of North America in Toronto.

As perhaps many of us know from direct experiment—best performed in a dark room in front of a mirror—wintergreen Lifesavers and many other hard candies made with granulated sugar emit little blue sparks when they are crunched. (The crunching can be carried out with pliers instead of the teeth.)

Grinding sugar crystals in a bowl also produces flashes of light, which are best seen in a dark room. Striking sugar cubes together, like striking a match, produces flashes of light, too. So does pulling sticky tape—adhesive tape, clear tape, masking tape—quickly from the roll or from a surface like glass.

Such phenomena, called triboluminescence, have been known since at least the 1600s, when Italian and English scientists described it. In England, for example, Robert Boyle once observed that "hard sugar being nimbly scraped with a knife would afford a sparkling light."

The question is, why?

Apparently, Dr. Sweeting explained in an interview, the fracturing of sugar crystals causes patches of positive and negative electrical charges to develop across the tiny cracks. When the electrical "pressure" between the charges becomes great enough, a discharge occurs.

It's the same principle that's involved with the little sparks that you emit when you touch a metallic doorknob after walking across a carpet on a cool, dry day. It's the same principle that's involved with lightning bolts. Positive and negative charges are fiercely attracted to one another, and when they're separated, they vigorously reunite with a flash of energy.

The tiny sparks that are emitted when you crunch a wintergreen or peppermint Lifesaver have, in fact, a lot in common with lightning, according to Dr. Sweeting. In both cases, the electrons—the carriers of negative electrical charges—bombard the nitrogen atoms in the air, causing them to emit blue-white light, which is the color of the sparks emitted by sugar that's being crunched.

If you would crunch sugar in an atmosphere of neon, which glows red when it's electrically charged, the sparks would be red.

Yet another variation of the phenomenon occurs with wintergreen Lifesavers. The oil of wintergreen flavoring itself emits light when it's energized by ultraviolet light, in the same sort of way, Dr. Sweeting explained, that ultraviolet light, or "black light," can make special paints or chalk glow.

The tiny lightning discharges that occur when wintergreen Lifesavers are crunched cause some of the nitrogen molecules' energy to be released in the ultraviolet range, and those ultraviolet rays cause the wintergreen flavoring's molecules themselves to emit light.

It's extremely difficult to study the light that wintergreen or peppermint Lifesavers and other materials give off when crunched, Dr. Sweeting noted, because the levels of the light are so low. The way scientists normally study light that's given off by things—from stars to atoms—is to direct the emitted light through special instruments that break it up into its component colors.

The light emitted from crunched sugar, however, is so weak that obtaining a spectrum the usual way is an arduous task. She noted that the first such spectrum was obtained in the 1930s by workers who ground sugar continuously in the dark for four to five hours.

Recently, Dr. Sweeting said, EG&G Princeton Applied Research developed a computerized image-intensifying device for analyzing low levels of light emitted by lasers. That was the device she used in her research on triboluminescence.

"That's the high-tech side of the study," she said. The low-tech side involved crunching particles of wintergreen Lifesavers, in a dark room, in a test tube covered by black cloth, all so that the instrument's little detector would see only the tiny lightning flashes from the Lifesavers.

Scientists still have a lot to learn about precisely what's going on in triboluminescence, Dr. Sweeting said. "At the moment, our ability to predict (which substances will have the property) is not very good," she noted.

Learning about the tiny lightning flashes from crunching Lifesavers may be more than an academic exercise.

One possible application, Dr. Sweeting suggested, could be in developing a remote sensing system for stresses and strains of vital components in, say, spacecraft or within nuclear reactors.

Materials, for example, could be coated or impregnated with triboluminescent substances, and if the material should begin developing tiny fractures or stresses, little flashes of light, detected by a fiber optics system, would provide an early warning system.

So there's potential value in learning about why we create tiny lightning strokes when we crunch a Lifesaver. Such studies are at the very least, you might say, enlightening.

Secret of the Gee-Haw-Whimmy-Diddle
OCTOBER 10, 1982

The tall bewhiskered man, dressed in bib overalls and wearing a Western straw hat, rubbed a stick across some notches whittled into a dowel, at the end of which was a propeller.

As he did so, the propeller spun like the blade of a tiny electric fan. If you stomp your foot, he told onlookers, the propeller will spin the other way. He stomped his foot and, as he continued stroking the notched stick, the propeller slowed, stopped and began turning in the opposite direction.

"Sometimes if you whistle, it will turn the other way," said the man. He whistled. The propeller reversed direction.

The man was one of a number of participants in the recent, annual arts and crafts show in New Market, where craftsmen displayed handiworks ranging from canes and toys to quilts, paintings and ingenious puzzles in metal and wood.

Of direct scientific interest was the notched propeller stick, also known variously as the magic propeller, a gee-haw-whimmy-diddle, a magic windmill or a hootie stick, presumably because of the claim that a roosting hoot owl's claws made the notches. The most common name applied to it seems to be gee-haw-whimmy-diddle.

Despite its general association with Southern mountain crafts, no one really seems to know the origin of the simple but engrossing toy. Roddy Moore, director of Ferrum College's Blue Ridge Institute, a rural life study center, suspects the gee-haw-whimmy-diddle has been associated with the mountains because some commercial suppliers have included it in their wares of Southern mountain crafts.

He said he does not know the origin of the device, and he knows of no credible authority in the field of folk craft who does know its history. But Moore mentioned one curious fact: "I have never seen an old one."

Questions of origin aside, it is an easily made device. Carve up to a dozen notches in a stick or dowel eight or so inches long; make a slightly oversized holed in a simple propeller and affix it with a think nail or carpet tack to the end of the stick. Some versions use square or rectangular sticks.

To make it work, one stokes another stick across the notches. But there is a secret. If you softly drag the side of your forefinger, on the rubbing stick, along one side the notched stick as you stroke it, the propeller will turn one way. If you place your thumb in contact with the other side of the notched stick during the stroking, the propeller will turn the other way.

Such subtle maneuvers may be imperceptible to onlookers, and therein lies the opportunity to concoct various farfetched explanations of the propeller's behavior. As put by Jearl Walker, author of the "Amateur Scientist" feature in *Scientific American* and a book called *The Flying Circus of Physics*, the number of lies that can be conjured to explain the propeller's behavior is almost limitless.

The craftsman at the New Market fair invoked foot-stomping and whistles as causes of the propeller's's reversals. Some have equated one's ability to cause the reversals with the amount of one's sex appeal. Walker blames it on cosmic rays. Others have used it as a device to answer questions; yes, for one direction and no for the other.

The major scientific questions about the device are, why does the propeller turn and why can it be made to switch directions? In his book, *The Flying Circus of Physics*, Walker cites five references in science journals to the behavior of the gee-haw-whimmy-diddle.

Most of those were in the form of some letters in the *American Journal of Physics* in the mid-1950s, and few contribute any substantial insight into the physics of the intriguing toy. The most serious attempt was by G. David Scott of the University of Toronto (his letter was in the Sept. 1956 issue of the *American Journal of Physics*.

He concluded that the vibrations set up in the stick, caused by rubbing a stick across the notches, causes the propeller's axle to vibrate. The axle vibrates in a line-like fashion with simple stroking of the stick. But dragging the forefinger or thumb on the side of the notched stick during the stroking process causes the axle to execute a circular motion, according to Scott.

He said the effect can be illustrated by punching a hole slightly larger than a pencil in a piece of cardboard. Place the pencil in the hole, execute a circular motion with the pencil and soon, a rotary motion will be imparted to the cardboard.

I am not sure that Scott's explanation is the complete one for the behavior of this intriguing toy. And I wonder how much of the explanation lies with the fact that nothing's quite perfect. The axle nail may not fit perfectly in the center of the stick; the hole in the propeller may not be perfectly circular.

And how much of a role does resonance play? Some empirical studies, for example, suggest that the spinning can be stopped if one chokes up too much of the notched stick. Also, it's possible to cause the propeller to spin without dragging a forefinger or thumb along the stick. Scott suggested that how one grasps the notched stick can influence the propeller. Pressing the stick at the 10 or 2 o'clock positions can make the propeller reverse directions, he wrote.

There is another aspect of the gee-haw-whimmy-diddle, however, that concerns me as much as the physics of the thing.

Who made the very first one and why?

Walk Or Run? The Better Way To Get Wetter
JULY 28, 1994

You are outside without an umbrella when a shower pops up. Suddenly you are faced with a problem—do you walk or run to the nearest shelter?

If you walk, you will be exposed to the rain longer than if you run, so you might get wetter than if you make a dash for it. But if you run you will collide with the raindrops all the way and, therefore, you might become wetter than if you walked.

What to do?

You could consider the theoretical analyses of the problem that some scientists have made over the years. One is by S.A. Stern of the University of Texas, who published a mathematical study that he titled, "An optimal speed for traversing a constant rain," in a 1983 issue of the *American Journal of Physics*.

Intuitively, he noted, you would think you should run as fast as you can so that you spend the least amount of time exposed to the rain.

But experience might suggest that a slower pace is better because the harder you run, the greater is the "apparent intensity" of the rain—the rain you run into.

Obviously, Stern reasoned, you can't just stand in the rain; you would become saturated if you did. So at least you should be in motion in the rain; the question is the speed at which you will become minimally wet. He tackled the problem by making some simplified assumptions.

He assumed that the rain was coming straight down. He used a square, inclined at various angles, as a substitute for the surface area that a person's head, shoulders and body would present to the vertically falling rain.

The conclusions he drew from his equations include:

• The less time you spend in the rain, or the faster you run, the better. Or as Stern put it, "Maximizing the velocity through the rain will be rainstrike minimizing."

• The number of raindrops hitting you vertically is not affected by how fast you run (remember, Stern's rain is "idealized rain;" it's rain that falls straight down).

• But the faster you run, the more raindrops you run into in the forward direction.

You can, of course, minimize the numbers of raindrops you run into by decreasing your forward surface area, the area that runs into the rain. You do that by leaning forward as much as possible.

So, according to Stern's analysis, the answer to the question of whether you should walk or run in the rain is that you should run. And you should lean forward as you run—if the rain is coming straight down. Taken to their logical extreme, Stern says his equations suggest that the best way to go through the rain is to lie prone on a skateboard and propel it as rapidly as possible.

In his book, *The Science of Everyday Life*, Jay Ingram cites another analysis of the problem, this one by an Italian physicist, Alessandro De Angelis, published in a 1987 issue of the *European Journal of Physics.* De Angelis also concludes that you will become less soaked if you run through the rain than if you walk through it.

But he also found that you become only slightly wetter if you walk briskly instead of running. De Angelis defined a brisk walk as being between six and seven miles per hour. Although you become less wet by running, De Angelis concludes that the small benefit you gain is not worth the extra effect; a brisk walk will do.

It's interesting that even without the benefit of such erudite studies, we tend to run or walk briskly when we're caught in the rain anyway.

Maybe we know more about scientific analysis than we realize, when we're put to the test.

Why Toast Hits Floor Jelly Down
NOVEMBER 30, 1995

We live in a universe in which the speed of light is 2.9979×10^8 meters per second; Avogadro's number is 6.025×10^{23} molecules per mole; and toast is prone to hit the floor jelly side down.

The first two are givens and are included in lists of fundamental constants that are on the back pages of many physics and chemistry textbooks. The item about the toast is a consequence of some basic principles.

Just how some of those principles conspire to make toast flip messy side down is described by Ian Stewart, in the "Mathematical Recreations" section of the December issue of *Scientific American,* based on analyses by a mathematically inclined British journalist and others.

As Stewart described it, the toast issue was raised several years ago through experiments performed by the host of a British Broadcasting Corp. television show. He tossed buttered toast in the air 300 times and found that it was no more likely to land butter side down than butter side up.

But the journalist, Robert Matthews, objected to the way the experiment was carried out. When toast falls on the floor, he said, it's not because we toss it in the air. Rather, it falls on the floor because it's accidentally knocked off the table.

Matthews developed a formula to describe how a slice of toast is most likely to flip as it falls from a table. [Matthews subsequently included his buttered toast studies in a wonderfully delightful article, "The Science of Murphy's Law," that appeared in the April 1997 issue of *Scientific American.*]

His formula shows that the slice will generally rotate about 180 degrees when knocked from a typical table, but not 360 degrees. If it flips 180 degrees, it will land butter side down. If it makes a complete 360 degree turn, it will land butter side up.

(One can verify this, as I have, by knocking a book off a table. If the front cover is up when you push it to the table's edge, the front cover will be down when it lands on the floor.)

The fact that there's butter on one side has little to do with it, Stewart noted. The butter adds no more than 10 percent to the total weight, he said, and much of it soaks within the toast toward the middle.

If a table were high enough, a piece of toast could make a complete 360 degree flip and land on its back, but in the real world, tables are not that high. As Stewart points out, table top heights are generally around or slightly lower than half the height of an average person.

For toast to do a 360 degree turn from a table top, Stewart said the table would have to be about 10 feet high, which would be an appropriate height for 20-foot-high people.

The toast-flipping issue may have universal implications. As Stewart noted, if there are creatures elsewhere in the universe who are something like us, and if they sit at tables and eat toast, their toast, too, is likely to flip butter side or jelly side down.

The reasons have to do with the ideal sizes of creatures. He referred to work by William H. Press of Harvard University who argued some years ago that creatures who walk on two legs can only grow so high. The gravitational field in which they live will restrict their heights.

Creatures that walk on two legs are more unstable than creatures that walk on four. And the maximum theoretical height for a two-legged creature is one at which it is not likely to suffer a fatal head injury if it should trip and fall.

Matthews figured, on theoretical grounds, that such a maximum height on Earth is 9 feet, 8 inches. Stewart noted that the tallest human on record was a man who was 8 feet, 11 inches tall.

And, he said, two-legged creatures on other planets would have about the same upper limit to their heights because of a complicated relationship between gravitational forces and the electrostatic forces that hold molecules together.

So the tables of even the largest bipedal aliens would not be high enough to prevent their toast from falling butter and jelly side down.

Some aspects of the universe just aren't pretty.

Movie Fiction Operates Under Different Rules
MARCH 22, 1987

Truth may often be stranger than fiction, but then again, fiction does have its moments, especially the fiction that's turned into the images of motion pictures.

There in the fiction land of movies, just as in the Wonderland and

Through-the-Looking-Glass places visited by Alice, rules different from those of our world sometimes apply.

Consider, for example, as has physicist Jack Weyland, some of the laws of physics. Sometimes, some of those laws are not strictly enforced in movie land, as Weyland pointed out a few months ago at a meeting in California.

Weyland teaches physics at the Technology South Dakota School of Mines and Technology, and he also gives popular lectures on physics to high school students in South Dakota, using segments from recent motion pictures to illustrate his points.

He gave a similar presentation in late January at a joint meeting of the American Physical Society and the American Association of Physics Teachers in San Francisco.

It must be quickly emphasized that Weyland does not intend to be a spoil-sport just because certain events in some movies violate laws of physics. In fact, he likes movies and occasional lapses in physical accuracy generally do not interfere with his enjoyment at all, he said in a telephone interview last week.

But as a physicist, here are some things that Weyland just couldn't help noticing:

• In a Superman movie, Lois Lane is falling from a tall building. She has fallen most of the way to the ground, when suddenly, Superman swoops up and catches her in his arms.

Considering the velocity of her falling body, and the upward velocity of Superman when his strong arms collide with her, Weyland believes Lois would be smashed more thoroughly than if she hit the pavement below.

And speaking of Superman, how exactly does his X-ray vision work? As Weyland notes, we see things because they emit or reflect light to our eyes, not because our eyes shoot out rays of light to objects we look at. If Superman does, in fact, send out X-rays to an object, the rays somehow must return to his eyes with the information they have acquired in order for him to sense it. But X-rays presumably go through the object and continue on.

• In *Star Wars*, Weyland wonders how laser beams, shooting through the vacuum of space, become so dazzlingly visible. We may see beams of light in our atmosphere, where dust, smoke and other tiny particles in the air make the beam visible, but in space, there's nothing.

• Also in *Star Wars*, fighter space craft bank and turn and engage in dog fights just as aircraft in our atmosphere do. But spacecraft are not flying through an atmosphere; they would not bank and turn and maneuver like aircraft because aerodynamic principles do not apply in space.

• A major puzzle in *Star Wars*—and in many other space movies—is the noise. There's the noise of spaceships, the noise of spaceships battling, and the noise of explosions. We perceive sound because some medium—air, usually—conveys it. There's no such thing as sound in a vacuum because there's nothing through which the energy can be transmitted.

• The command, "Beam me up, Scotty," from *Star Trek* movies sent Weyland to his calculator. Presumably, Enterprize crew members who are being beamed up or down are converted into energy, transmitted, then converted into matter again.

Using Einstein's famous equation, which says that energy is equal to mass times the square of the speed of light, Weyland figured that the energy in, say, Mr. Spock's body would be more than 1,000-fold greater than that released by a 1-megaton thermonuclear bomb. ("Stand back—way, way, way back—and beam me up, Scotty.")

• In *Raiders of the Lost Ark*, Indiana Jones deftly exchanges a bag of sand, held in one hand, with gold figure that rests on a specially rigged platform. The weight of the gold figures prevents devices from being set off that would kill the intruder. Indiana Jones quickly sweeps away the gold figure and replaces it with a sand bag of about the same size as the idol.

But, notes Weyland, the weights could not have been the same. Gold is much more dense than sand. The approximately one-foot-tall figure would weigh at least 60 pounds, according to Weyland. A 60-pound bag of sand would be considerably larger, and certainly could not be handled deftly with one hand, according to Weyland, who concludes that the idol must really have been a plastic one.

• In a Ninja martial arts movie, a person springs from the floor upward with such force that he crashes through the ceiling and through the floor of the upstairs room. In order to propel himself upward with such force, Weyland said the man would have to push down on the floor he was standing on with a force at least equal to that required to crash through the ceiling. That's all according to Newton's third law of motion.

Thus, according to Weyland, the man is more likely to crash through the floor he's standing on than to shoot upward with enough force to break through the ceiling and floor above.

Such violations of physical law bother him a bit, but for the most part, Weyland says he enjoys the movies, lapses and all. He loved *Star Wars*, for example.

He said that one part of his mind was taking note of the physics violations.

"But one part was saying, "'Who cares?'"

It's Only a Romano Moon
AUGUST 16, 1970

Of what is the moon made?

More than a year after man's first landing there, and after many months of probing and prodding of moon rocks by hundreds of Earth scientists, one lesson is emerging—the moon's rocks are different, by and large from Earth rocks.

But, guided by an old hypothesis about the nature of the moon, a team of geologists has found a group of Earth materials that do have certain properties in common with the lunar rocks.

The Earth materials are cheeses.

Or so the geologists report in a recent spread of charts, graphs and technical data in the journal *Science*. Their report has few holes in it, even for an admittedly tongue-in-cheek project.

The starter in this case was an observation that sound waves travel more slowly through a couple of samples of moon rocks than through known Earth rocks.

Among those noting this difference between lunar rocks and Earth material were two geologists at the Lamont-Doherty Geological Observatory at Palisades, N.Y., Dr. Edward Schreiber and Dr. Orson L. Anderson.

They and their colleagues had discovered the "startlingly low" velocities of sound waves through two lunar rock samples by measuring the time it took for pulses to be transmitted through known thicknesses of the samples.

In one sample, the sound waves traveled at a velocity of 11.1 miles per second. In the other, it was 0.78 miles per second. (The approximate speed of sound through the air is about 0.22 miles per second.) Further, the results of the Apollo 12 seismic experiment, in which the lunar module was deliberately crashed on the moon to produce a pulse that was recorded by a seismograph, agrees well with the laboratory studies of the lunar samples.

Earthbound rocks and minerals, on the other hand, transmit sound waves much more quickly, as indicated in a chart that accompanied the article by Drs. Schreiber and Anderson.

For sedimentary rocks such as limestone and sandstone, for example, sound velocities are around three miles a second. For metamorphic rocks, such as slate and marble, it ranges from three to more than four miles a second. for igneous rocks such as granite, it is around four miles a second, and for various minerals, including quartz and garnet, it ranges from about five to nearly seven miles a second.

Hence, the puzzle—and the beginning of the tongue-in-cheek project of Dr. Schreiber and Dr. Anderson.

Twelve years ago, while he was working toward his Ph.D., Dr. Schreiber once had to answer an examination question that asked for the tests that could be performed to show that the "moon was or was not made of cheese."

The idea of cheese's relation to the moon had been fermenting at the back of his mind ever since, he indicated in a telephone interview last week.

Thus, when the unusual sound transmission properties of the lunar rocks were discovered, and the usual Earth rocks and minerals did not compare well with these properties, "I just felt in my bones that cheese would fit the bill," said Dr. Schreiber.

So he and Dr. Anderson actually began testing cheeses in the same way that the lunar samples were studied. They sent mechanically produced pulses through various cheeses, and measured the transit times of those pulses through known thicknesses of the cheeses.

"The materials studied," they soberly stated in their article, "were chosen so as to represent a broad geographic distribution in order to preclude any bias that might be introduced by regional sampling."

The results—the velocities of sound waves through the cheeses—clustered around the values found for the lunar samples.

Gjetost of Norway, for example, transmitted sound at a velocity of 1.1 miles per second. Sapsego (Swiss), 1.3 miles per second. Munster of Wisconsin, 0.97 miles per second. Provolone and Romano of Italy, 1.09 miles per second. Cheddar (Vermont), 1.72 miles a second.

In the largely straight-faced journal article, the authors, noting that previous searches failed to find Earth analogies of lunar material, said their search "was aided by considerations of much earlier speculations concerning the nature of the moon." A footnote referred to a quotation from Erasmus:

"With this pleasant merry toy, he...made his friends believe the moon to be made of green cheese."

To this, Drs. Schreiber and Anderson added, "It is seen that these materials (cheeses) exhibit compressional velocities that are in consonance with those measured for the lunar rocks—which leads us to suspect that perhaps old hypotheses are best, after all, and should not be lightly discarded."

Since their paper appeared in the June 26, 1970 issue of *Science*, the two geologists have received a number of applauding letters from scientists around the country. One was addressed to "Lunatic Edward Schreiber." Another extended their work by suggesting the development of a theory of lunar maria formation "on the basis of the toasted cheese sandwich...."

For all the fun they have had in their project, the geologists are now astonished to learn that their project may have a wider legitimacy after all.

Cheeses have a lot of water in them; their project is beginning to teach them "how to deal with water in minerals; how to take water into account in the relationship of velocity to density," said Dr. Schreiber.

He didn't suggest the wider implications of cheese, moon and water. Perhaps someday, a well-aged songwriter might pen: "It's only a Romano moon, shining on a sapsago sea...."

What's Meant to Be Is Often Full of Paradox
DECEMBER 27, 1993

A week from today we will be in 1994, the top of a new, year-long hill in time—whatever time is.

Actually, some of the people who have thought the most about time are not scientists but the fiction writers who dream up stories about time machines. Modern examples include the script writers for the *Back to the Future* movie trilogy.

Such writers have to think about time because a successful time machine immediately raises, among other things, issues about the paradoxes of meddling.

If you go back in time and somehow cause your great-grandfather's death, then you could not exist today to go back in time to cause your great-grandfather's death.

Or, if you travel into the future and carve your initials in a particular tree, does that mean that no one nor any force can destroy that tree between now and then? It will, after all, have to be there in the future for you to carve your initials on it.

Because of such paradoxes, the time machine fiction writers often seem to conclude that it's best not to do anything that could alter the present, whether you are visiting the past or the future.

And behind their apparent conclusion is a slant on one of the oldest pair of philosophical questions there is about time. In the context of the New Year season, the questions may be put like this:

Are the things that are going to happen during the coming year already set by some cosmic script? Are events in the stars meant to be, merely to unfold as time passes?

Or is there no script, no destiny, with the future mostly to be determined as it comes?

Those who lean to the "meant-to-be" philosophy draw some support from science. The motions of the planets around the sun are known with such clarity, as are the forces that govern them, that their future positions can be predicted with great precision.

Scientists can tell us the day, hour, minute and second that an eclipse of the sun or moon will occur next year or next century. They can tell, with similar accuracy, when such eclipses occurred in the past.

They can predict the phases of our moon; sunsets and sunrises, the positions of Jupiter's moons and the future fate of our sun.

All such predictions are possible because those things are, in a sense, meant to be; it takes only the unfolding of time for the events to be played out according to the well-known laws of celestial mechanics.

And some scientists have claimed, in the past anyway, that if they knew enough about the present positions and motions of all the atoms in the universe, they could write out the entire future of our universe, suggesting that all future events are already determined.

But those who lean to an "undetermined" theory of the future also draw some support from science. Take complex events like the weather, for instance.

Sure, you can make reasonable predictions in the short run, the nondeterminists might say, but tiny uncertainties will quickly build up and throw your predictions off in a week or less.

Even if you knew the temperatures and all other measures of the atmosphere for every cubic foot of air from the surface to 50 miles up, it's those uncertainties down at the third, fourth, fifth or sixth decimal places of the measurements that will rapidly grow and overwhelm your predictions.

The world is simply too complex, too full of unknowables, too dependent on undetermined, chancy encounters to have a tightly scripted future, the nondeterminists might argue.

Interestingly, the *Back to the Future* script writers have allowed their characters to meddle slightly with the past to produce certain outcomes in the present. And in their final movie, *Back to the Future Part III*, they allow history to be changed.

Doc, the time-machine scientist, prevents a woman from the past from being killed—even though the cliff over which she was supposed to have plunged was known by her name a century later because of her fatal accident. Doc married that woman from the past and they had two children (named Jules and Verne, incidentally, for the master writer about time machines).

When one of Doc's fellow time travelers questioned him at the end of the movie about a certain future event, he replied—obviously abandoning any belief in predestined events:

"Your future hasn't been written yet. No one's has."

Then he added what could be an appropriate New Year's sentiment:

"Your future is what you make it, so make it a good one."

W

Y

Z

www.ingramcontent.com/pod-product-compliance
Lightning Source LLC
Chambersburg PA
CBHW060023210326
41520CB00009B/977